CliffsStudySolver™

Trigonometry

CliffsStudySolver™
Trigonometry

By David Alan Herzog

WILEY

Wiley Publishing, Inc.

Published by:
Wiley Publishing, Inc.
111 River Street
Hoboken, NJ 07030-5774
www.wiley.com

Copyright © 2005 Wiley, Hoboken, NJ

Published by Wiley, Hoboken, NJ
Published simultaneously in Canada

Library of Congress Cataloging-in-Publication Data

Herzog, David Alan.
    Trigonometry / by David A. Herzog.
        p. cm. -- (CliffsStudySolver)
    Includes index.
    ISBN-10: 0-7645-7968-1 (pbk.)
    ISBN-13: 978-0-7645-7968-4
    1. Trigonometry--Problems, exercises, etc. I. Title. II. Series.
    QA537.H47 2005
    516.24--dc22

                                            2005007454

Printed in the United States of America

10 9 8 7 6 5 4 3 2 1

1B/RR/QV/QV/IN

For general information on our other products and services or to obtain technical support, please contact our Customer Care Department within the U.S. at 800-762-2974, outside the U.S. at 317-572-3993, or fax 317-572-4002.

Wiley also publishes its books in a variety of electronic formats. Some content that appears in print may not be available in electronic books. For more information about Wiley products, please visit our web site at www.wiley.com.

About the Author

David Alan Herzog is the author of numerous books concerned with test preparation in mathematics and science. Additionally, he has authored over one hundred educational software programs. Prior to devoting his full energies to authoring educational books and software, he taught math education at Fairleigh Dickinson University and William Paterson College, was mathematics coordinator for New Jersey's Rockaway Township Public Schools, and taught in the New York City Public Schools.

Dedication

This book is dedicated to Francesco, Sebastian, and Gino Nicholas Bubba, Rocio, Kira, Jakob and Myles Herzog, Hailee Foster, all of their parents, and Uncles Dylan and Ian.

Publisher's Acknowledgments

Editorial

Project Editor: Suzanne Snyder

Acquisitions Editor: Greg Tubach

Technical Editor: Tom Page

Editorial Assistant: Meagan Burger

Composition

Project Coordinator: Ryan Steffen

Proofreader: Evelyn Still

Indexer: Ty Koontz

Wiley Publishing, Inc. Composition Services

Table of Contents

Trigonometry Pretest   1
    Answers   16
Chapter 1: Trigonometric Ideas   21
  Angles and Quadrants   21
  Coterminal Angles   25
  Trigonometric Functions of Acute Angles   28
    Reciprocal Trigonometric Functions   29
  Introducing Trigonometric Identities   32
    Trigonometric Cofunctions   34
    Two Special Triangles   35
  Functions of General Angles   38
    Reference Angles   41
    Squiggly versus Straight   43
  Trig Tables versus Calculators   44
  Interpolation   45
  Chapter Problems and Solutions   49
    Problems   49
    Answers and Solutions   51
  Supplemental Chapter Problems   55
    Problems   55
    Answers   57
Chapter 2: Graphs of Trigonometric Functions   59
  Understanding Degree Measure   59
  Understanding Radians   59
  Relationships between Degrees and Radians   60
  The Unit Circle and Circular Functions   62
  Domain versus Range   63
  Periodic Functions   66
  Graphing Sine and Cosine   67
    Vertical Displacement and Amplitude   70
    Frequency and Phase Shift   74
    Graphing Tangents   77
    Asymptotes   77
    Graphing the Reciprocal Functions   78

Chapter Problems and Solutions   84
    Problems   84
    Answers and Solutions   85
Supplemental Chapter Problems   89
    Problems   89
    Answers   90

Chapter 3: Trigonometry of Triangles   93
Finding Missing Parts of Right Triangles   93
Angles of Elevation and Depression   95
The Law of Sines   100
The Law of Cosines   105
Solving General Triangles   108
    SSS   109
    SAS   109
    ASA   111
    SAA   111
    SSA, The Ambiguous Case   111
Areas of Triangles   116
    Area for SAS   116
    Area for ASA or SAA   117
    Heron's Formula (SSS)   119
Chapter Problems and Solutions   122
    Problems   122
    Answers and Solutions   124
Supplemental Chapter Problems   133
    Problems   133
    Answers   134

Chapter 4: Trigonometric Identities   137
Fundamental Identities   137
    Reciprocal Identities   137
    Ratio Identities   138
    Cofunction Identities   139
    Identities for Negatives   139
    Pythagorean Identities   140
Addition and Subtraction Identities   143
Double Angle Identities   144
Half Angle Identities   146
Tangent Identities   150

Product-Sum and Sum-Product Identities   152
   Product-Sum Identities   152
   Sum-Product Identities   153
Chapter Problems and Solutions   155
   Problems   155
   Answers and Solutions   156
Supplemental Chapter Problems   163
   Problems   163
   Answers   164

Chapter 5: Vectors   167
Vectors versus Scalars   167
Vector Addition Triangle/The Tip-Tail Rule   169
Parallelogram of Forces   172
Vectors in the Rectangular Coordinate System   178
Resolution of Vectors   180
Algebraic Addition of Vectors   183
Scalar Multiplication   185
Dot Products   185
Chapter Problems and Solutions   188
   Problems   188
   Answers and Solutions   190
Supplemental Chapter Problems   198
   Problems   198
   Answers   201

Chapter 6: Polar Coordinates and Complex Numbers   203
Polar Coordinates   203
Converting between Polar and Rectangular Coordinates   206
   Converting from Polar to Rectangular Coordinates   206
   Converting from Rectangular to Polar Coordinates   206
Some Showy Polar Graphs   211
Plotting Complex Numbers on Rectangular Axes   217
Plotting Complex Numbers on the Polar Axis   219
Conjugates of Complex Numbers   223
Multiplying and Dividing Complex Numbers   224
Finding Powers of Complex Numbers   225
Chapter Problems and Solutions   228
   Problems   228
   Answers and Solutions   230

Supplemental Chapter Problems 237

    Problems 237

    Answers 239

Chapter 7: Inverse Functions and Equations 241

Restricting Functions 242

Inverse Sine and Cosine 244

Inverse Tangent 249

Inverses of Reciprocal Functions 253

Trigonometric Equations 257

Uniform Circular Motion 261

Simple Harmonic Motion 263

Chapter Problems and Solutions 266

    Problems 266

    Answers and Solutions 268

Supplemental Chapter Problems 272

    Problems 272

    Answers 273

Customized Full-Length Exam 275

Problems 275

Appendix A: Summary of Formulas 293

Basic Trigonometric Functions 293

Reciprocal Identities 293

Ratio Identities 294

Trigonometric Cofunctions 294

Identities for Negatives 294

Pythagorean Identities 294

Opposite Angle Identities 295

Double Angle Identities 295

Half Angle Identities 295

Sum and Difference Identities 295

Product-Sum Identities 296

Sum-Product Identities 296

Inverse Identities 296

Appendix B: Trigonometric Functions Table 297

Glossary 301

Index 305

Trigonometry Pretest

Directions: Questions 1 through 75.

Where it appears, the symbol ∠ stands for "angle"; ∠s is its plural. You will need either a scientific calculator or the table of trigonometric functions on page 297 to answer certain questions. Where appropriate, approximate the value of π as 3.14.

Circle the letter of the appropriate answer.

1. In which quadrant does a 75° angle in standard position have its terminal side?
   A. I
   B. II
   C. III
   D. IV

2. In which quadrant does a 175° angle in standard position have its terminal side?
   A. I
   B. II
   C. III
   D. IV

3. In which quadrant does a 375° angle in standard position have its terminal side?
   A. I
   B. II
   C. III
   D. IV

4. Which angle is coterminal with a 45° angle in standard position?
   A. 225°
   B. 295°
   C. 425°
   D. 765°

5. Which angle is coterminal with a 125° angle in standard position?

   A. 205°

   B. 375°

   C. 485°

   D. 665°

6. Which of the following ratios gives the sine of an angle in standard position?

   A. $\dfrac{\text{opposite}}{\text{adjacent}}$

   B. $\dfrac{\text{hypotenuse}}{\text{opposite}}$

   C. $\dfrac{\text{opposite}}{\text{hypotenuse}}$

   D. $\dfrac{\text{adjacent}}{\text{hypotenuse}}$

7. Which of the following ratios gives the tangent of an angle in standard position?

   A. $\dfrac{\text{opposite}}{\text{adjacent}}$

   B. $\dfrac{\text{hypotenuse}}{\text{opposite}}$

   C. $\dfrac{\text{opposite}}{\text{hypotenuse}}$

   D. $\dfrac{\text{adjacent}}{\text{hypotenuse}}$

8. Which of the following ratios gives the cosecant of an angle in standard position?

   A. $\dfrac{\text{opposite}}{\text{adjacent}}$

   B. $\dfrac{\text{hypotenuse}}{\text{opposite}}$

   C. $\dfrac{\text{opposite}}{\text{hypotenuse}}$

   D. $\dfrac{\text{adjacent}}{\text{hypotenuse}}$

9. In which quadrants is sine function negative?

   A. I and II

   B. I and III

   C. II and IV

   D. III and IV

10. In which quadrants is tangent function negative?

    A. I and II

    B. I and III

    C. II and IV

    D. III and IV

11. Given tan 28.40° = 0.5407 and tan 28.50 = 0.5430, what is tan 28.43?

    A. 0.5414

    B. 0.5417

    C. 0.5420

    D. 0.5424

12. Cos 70° = 0.3420 and cos 71° = 0.3256. What is the cosine of 70.6°?

    A. 0.3387

    B. 0.3355

    C. 0.3322

    D. 0.3289

13. What is the degree measure of a circle's central angle that subtends an arc that is $\frac{1}{3}$ the length of the circumference?

    A. 90°

    B. 120°

    C. 150°

    D. 180°

14. What is the radian equivalent of the angle measure of a 90° angle?

    A. $\frac{\pi}{4}$

    B. $\frac{\pi}{2}$

    C. π

    D. 2π

15. What is the degree equivalent of the angle measure of $\frac{2\pi}{3}$ radians?

    A. 100°

    B. 110°

    C. 120°

    D. 130°

16. Given a unit circle with a right triangle drawn in standard position, the angle at the origin named θ, and the hypotenuse being the radius of the circle, which of the following names the coordinates of the terminal side of the central angle?

    A. $(\cos\theta, \tan\theta)$

    B. $(\sin\theta, \cos\theta)$

    C. $(\tan\theta, \sin\theta)$

    D. $(\cos\theta, \sin\theta)$

17. What is the domain of the sine function?

    A. the set of real numbers

    B. the set of positive numbers

    C. the set of negative numbers

    D. all numbers from negative one to positive one

18. What is the range of the cosine function?

    A. the set of real numbers

    B. the set of positive numbers

    C. the set of negative numbers

    D. all numbers from negative one to positive one

19. What is the period of an 850° angle?

    A. 90°

    B. 110°

    C. 130°

    D. 150°

20. What is the period of an angle of 7π radians?

    A. $\frac{\pi}{2}$

    B. π

    C. $\frac{3\pi}{2}$

    D. 2π

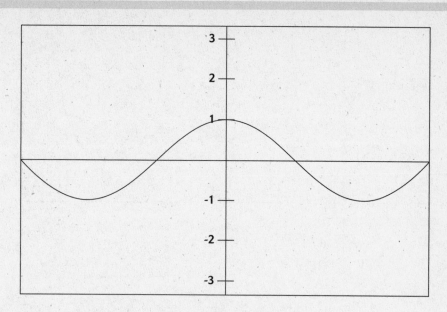

21. Which trigonometric function does the graph illustrate?

    A. $y = \sin x$
    B. $y = \cos x$
    C. $y = \tan x$
    D. $y = \sec x$

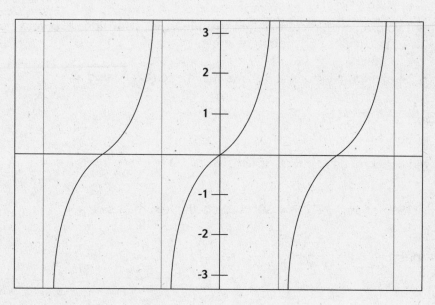

22. Which trigonometric function does the graph illustrate?

    A. $y = \sin x$
    B. $y = \cos x$
    C. $y = \tan x$
    D. $y = \sec x$

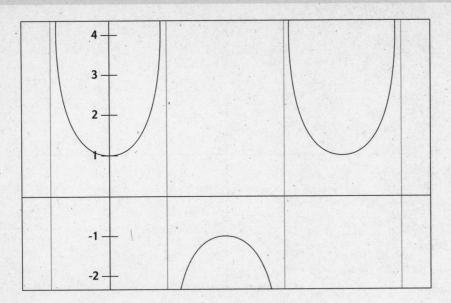

23. Which trigonometric function does the graph illustrate?

A. $y = \sin x$

B. $y = \cos x$

C. $y = \tan x$

D. $y = \sec x$

For the following two problems, picture a right triangle, *ABC*, with sides *a*, *b*, and *c* opposite ∠A, ∠B, and ∠C, respectively. ∠C is the right angle.

24. If side *c* is 24 mm long and ∠A = 30°, what is the length of side *a*?

A. 12 mm

B. 14 mm

C. 16 mm

D. 18 mm

25. If side *c* is 18 mm long and ∠B = 30°, what is the length of side *a*?

A. 9.2 mm

B. 11.4 mm

C. 13.5 mm

D. 15.6 mm

26. Bill is 12 feet away from the foot of a tree. Lying on his belly and sighting up from the ground along a protractor, Bill finds the treetop to form a 50° angle of elevation with the ground. To the nearest foot, how tall is the tree?

A. 12 ft

B. 14 ft

C. 16 ft

D. 18 ft

27. An airplane's radar shows its target to be 30 miles away. The angle of depression of the target is 21°. To the nearest tenth of a mile, what is the airplane's altitude?

A. 8.8 miles

B. 9.8 miles

C. 10.8 miles

D. 11.8 miles

28. In $\angle QRS$, $\angle Q = 35°$, while $\angle R = 65°$. Side q is 12 cm. To the nearest cm, find the length of side s.

A. 15 cm

B. 18 cm

C. 21 cm

D. 24 cm

29. If the angles of a triangle are α, β, and γ, and a, b, and c are the respective sides opposite those angles, then if $a = 6$, $b = 7$, and $c = 12$, find the measure of β to the nearest degree.

A. 20°

B. 25°

C. 30°

D. 35°

30. If all three sides of a triangle are known, how do we find the measure of the first angle?

A. The Law of Sines

B. The Law of Cosines

C. Heron's Formula

D. Pythagorean Identities

31. If two sides of a triangle and their included angle are known, how do we find the measure of the third side?

    A. The Law of Cosines

    B. The Law of Sines

    C. Heron's Formula

    D. Pythagorean Identities

32. If two angles of a triangle and their included side are known, we can subtract from 180° to find the third angle. Then how do we find the measures of the other two sides?

    A. The Law of Cosines

    B. The Law of Sines

    C. Heron's Formula

    D. Pythagorean Identities

33. If two angles of a triangle and a nonincluded side are known, we can subtract from 180° to find the third angle. Then how do we find the measures of the other two sides?

    A. The Law of Sines

    B. The Law of Cosines

    C. Heron's Formula

    D. Pythagorean Identities

34. If two sides of a triangle and a nonincluded angle are known, it is known as "The Ambiguous Case." Why is that so?

    A. There are two possible solutions.

    B. There are three possible solutions.

    C. There are four possible solutions.

    D. There are six possible solutions.

35. One of the following formulas may be used to find the area of a triangle. Which one?

    A. $K = \frac{1}{2} ab \sin B$

    B. $K = \frac{1}{2} bc \sin B$

    C. $K = \frac{1}{2} ac \sin B$

    D. $K = \frac{1}{2} ab \sin A$

36. Which of the following names describes the formula $K = \sqrt{s(s-a)(s-b)(s-c)}$?

    A. The Law of Sines

    B. The Law of Cosines

    C. Heron's Formula

    D. A Pythagorean Identity

37. What is the area of a triangle whose sides measure 10, 14, and 20 inches (to the nearest inch)?

    A. 55 in^2

    B. 65 in^2

    C. 75 in^2

    D. 85 in^2

38. Which of the following represents the reciprocal identity for sine?

    A. $\sin\theta = \dfrac{1}{\cos\theta}$

    B. $\sin\theta = \dfrac{1}{\sec\theta}$

    C. $\sin\theta = \dfrac{1}{\csc\theta}$

    D. $\sin\theta = \dfrac{1}{\tan\theta}$

39. Which of the following represents the reciprocal identity for cosine?

    A. $\cos\theta = \dfrac{1}{\sin\theta}$

    B. $\cos\theta = \dfrac{1}{\sec\theta}$

    C. $\cos\theta = \dfrac{1}{\csc\theta}$

    D. $\cos\theta = \dfrac{1}{\tan\theta}$

40. Which of the following ratio identities is true?

    A. $\tan\phi = \dfrac{\cos\phi}{\sin\phi}$

    B. $\cos\phi = \dfrac{\tan\phi}{\sin\phi}$

    C. $\sin\phi = \dfrac{\cos\phi}{\tan\phi}$

    D. $\tan\phi = \dfrac{\sin\phi}{\cos\phi}$

41. Which of the following ratio identities is true?

    A. $\cot\phi = \dfrac{\cos\phi}{\sin\phi}$

    B. $\cos\phi = \dfrac{\tan\phi}{\sin\phi}$

    C. $\sec\phi = \dfrac{\sin\phi}{\tan\phi}$

    D. $\cot\phi = \dfrac{\sin\phi}{\cos\phi}$

42. Which of the following cofunction identities is true?

    A. $\sin\gamma = \cos(45° - \gamma)$

    B. $\sin\gamma = \cos(90° - \gamma)$

    C. $\sin\gamma = \cos(180° - \gamma)$

    D. $\sin\gamma = \cos(360° - \gamma)$

43. Which of the following Pythagorean identities is true?

    A. $\tan^2\alpha + 1 = \cos^2\alpha$

    B. $\cos^2\alpha + 1 = \sec^2\alpha$

    C. $\tan^2\alpha + 1 = \sec^2\alpha$

    D. $\cot^2\alpha + 1 = \sec^2\alpha$

44. Which of the following identities for negatives is true?

    A. $\sin(-\theta) = -\sin\theta$

    B. $\sin(-\theta) = -\cos\theta$

    C. $\sin(-\theta) = -\sec\theta$

    D. $\sin(-\theta) = -\csc\theta$

45. What is the trigonometric identity for $\cos(\alpha + \beta)$?

    A. $\cos(\alpha + \beta) = \cos\alpha\cos\beta + \sin\alpha\sin\beta$

    B. $\cos(\alpha + \beta) = \cos\alpha\cos\beta - \sin\alpha\sin\beta$

    C. $\cos(\alpha + \beta) = \sin\alpha\cos\beta + \cos\alpha\cos\beta$

    D. $\cos(\alpha + \beta) = \sin\alpha\cos\beta - \sin\alpha\sin\beta$

46. What is the trigonometric identity for $\cos(\alpha - \beta)$?

    A. $\sin(\alpha - \beta) = \sin\alpha\cos\beta + \cos\alpha\sin\beta$

    B. $\sin(\alpha - \beta) = \sin\alpha\sin\beta - \cos\alpha\sin\beta$

    C. $\sin(\alpha - \beta) = \sin\alpha\cos\beta - \cos\alpha\sin\beta$

    D. $\sin(\alpha - \beta) = \sin\alpha\cos\beta - \cos\alpha\cos\beta$

47. Which of the following is *not* true of $\sin2\phi$?

    A. $\sin2\phi = 2\sin\phi + \cos\phi$

    B. $\sin2\phi = \sin\phi\cos\phi + \cos\phi\sin\phi$

    C. $\sin2\phi = \sin(\phi + \phi)$

    D. $\sin2\phi = 2\sin\phi\cos\phi$

48. Which of the choices is correct for the identity of $\cos 2\theta$?

      I. $\cos 2\theta = \cos^2\theta - \sin^2\theta$

      II. $\cos 2\theta = 2\cos^2\theta - 1$

      III. $\cos 2\theta = 1 - 2\sin^2\theta$

A. I and II only

B. II only

C. I and III only

D. I, II, and III

49. Which of these choices is *not* a correct half angle identity?

A. $\tan\dfrac{\theta}{2} = \dfrac{\sin\theta}{1 - \cos\theta}$

B. $\cos\dfrac{\theta}{2} = \pm\sqrt{\dfrac{1 + \cos\theta}{2}}$

C. $\sin\dfrac{\theta}{2} = \pm\sqrt{\dfrac{1 - \cos\theta}{2}}$

D. $\tan\dfrac{\theta}{2} = \dfrac{1 - \cos\theta}{\sin\theta}$

50. Which is the trigonometric identity for $\tan(\alpha + \beta)$?

A. $\tan(\alpha + \beta) = \dfrac{\tan\alpha - \tan\beta}{1 + \tan\alpha\tan\beta}$

B. $\tan(\alpha + \beta) = \dfrac{\tan\alpha + \tan\beta}{1 - \tan\alpha\tan\beta}$

C. $\tan(\alpha + \beta) = \dfrac{\tan\alpha + \tan\beta}{1 + \tan\alpha\tan\beta}$

D. $\tan(\alpha + \beta) = \dfrac{\tan\alpha - \tan\beta}{1 - \tan\alpha\tan\beta}$

51. Which of the choices is correct for the identity of $\dfrac{\lambda}{2}$?

      I. $\tan\dfrac{\lambda}{2} = \dfrac{\sin\lambda}{1 + \cos\lambda}$

      II. $\tan\dfrac{\lambda}{2} = \dfrac{1 - \cos\lambda}{\sin\lambda}$

      III. $\tan\dfrac{\lambda}{2} = \pm\sqrt{\dfrac{1 - \cos\lambda}{1 + \cos\lambda}}$

A. I and II only

B. II only

C. I and III only

D. I, II, and III

52. Which of the following is *not* a product-sum identity?

  A. $\sin \alpha \cos \beta = \frac{1}{2}\left[\sin(\alpha + \beta) + \sin(\alpha - \beta)\right]$

  B. $\sin \alpha \cos \beta = \frac{1}{2}\left[\cos(\alpha - \beta) + \cos(\alpha - \beta)\right]$

  C. $\sin \alpha \cos \beta = \frac{1}{2}\left[\sin(\alpha - \beta) + \sin(\alpha + \beta)\right]$

  D. $\sin \alpha \cos \beta = \frac{1}{2}\left[\sin(\alpha - \beta) + \sin(\alpha + \beta)\right]$

53. What is the name given to a speed or force that also has direction?

  A. scalar

  B. magnitude

  C. vector

  D. norm

54. Two vectors are arranged tip to tail so as to form an obtuse angle. The other ends of the two are connected so as to form an obtuse triangle. What is the sum of the vectors?

  A. the sum of the magnitudes of the two original vectors

  B. the side of the triangle opposite the obtuse angle

  C. the sum of the two short sides of the triangle

  D. none of the above

Questions 55–57 refer to the same airplane.

55. An airplane is traveling due west at an airspeed of 500 mph. A 75 mph tailwind is blowing from the northeast. What figure can best be used to determine the plane's final course and groundspeed?

  A. triangle

  B. rectangle

  C. parallelogram

  D. trapezoid

56. In question 55 the bearing of the wind is 210°. What is the plane's true course (0° is true north) to the nearest degree?

  A. 260°

  B. 263°

  C. 266°

  D. 269°

57.   What is the plane's groundspeed to the nearest mph?

   A. 541 mph

   B. 551 mph

   C. 561 mph

   D. 571 mph

58.   Vector $\overrightarrow{AB}$ has coordinates of A (-3, -5) and B (5, 8). What are the coordinates of point P such that $\overrightarrow{OP}$ is a standard vector and $\overrightarrow{OP} = \overrightarrow{AB}$?

   A. (2, 7)

   B. (4, 9)

   C. (6, 11)

   D. (8, 13)

59.   What is the magnitude of vector **v** = (3, -5)?

   A. $\sqrt{-8}$

   B. $2\sqrt{2}$

   C. $4\sqrt{3}$

   D. $\sqrt{34}$

60.   If vector **v** = (5, 7) and vector **u** = (6, -5), what is the sum of **v** + **u**?

   A. (0, 13)

   B. (11, 2)

   C. (12, 1)

   D. (13, 0)

61.   If **v** = (6, -3) and **u** = (5, 8) then what is 5**v** $-$ 3**u**?

   A. (-15, -39)

   B. (15, -39)

   C. (45, 39)

   D. (45, -39)

62.   Write **v** = (6, -3) in terms of the unit vectors **i** and **j**.

   A. 6**i** $-$ 3**j**

   B. 3(**j** $-$ 2**i**)

   C. 3(**i** $-$ **j**)

   D. 6**j** $-$ 3**i**

63. Find **5u** + **6v** if **u** = 6**i** − 4**j** and **v** = −3**i** + 7**j**.

    A. 6**i** + 11**j**

    B. 2(6**i** + 11)**j**

    C. 12**i** − 22**j**

    D. −12**i** − 22**j**

64. What are the polar coordinates for the point $P(5, 8)$?

    A. $\left(\sqrt{85}, 54°\right)$

    B. $\left(\sqrt{87}, 56°\right)$

    C. $\left(\sqrt{89}, 58°\right)$

    D. $\left(\sqrt{91}, 60°\right)$

65. What are the rectangular coordinates for the point $P(8, 30°)$?

    A. (3.6, 2)

    B. (4.7, 2.5)

    C. (5.8, 3)

    D. (6.9, 4)

66. Transform the equation $x^2 + y^2 + 7x = 0$ into polar coordinates form.

    A. $r + 7\cos\theta = 0$

    B. $r + 7\sin\theta = 0$

    C. $r + 7\tan\theta = 0$

    D. $r + 7\csc\theta = 0$

67. On the complex plane, in which quadrant would the graph of $-5 - 3i$ appear?

    A. I

    B. II

    C. III

    D. IV

68. By which pair of equations can complex numbers be converted to polar coordinates?

    A. $x = r\sin\theta, y = r\cos\theta$

    B. $x = r\cos\theta, y = r\sin\theta$

    C. $r = x\cos\theta, y = r\sec\theta$

    D. $x = r\sin\theta, y = r\csc\theta$

69. According to DeMoivre's Theorem, which equation represents a true statement?

    A. $z^2 = r^2(\cos 2\phi + i \sin 2\phi)$

    B. $z^3 = r^3(\cos 3\phi + i \sin 3\phi)$

    C. $z^4 = r^4(\cos 4\phi + i \sin 4\phi)$

    D. all of the above

70. What are the restrictions on the range of the inverse sine function?

    A. $-2\pi \leq y \leq 0$

    B. $-\pi \leq y \leq 0$

    C. $0 \leq y \leq \pi$

    D. $0 \leq y \leq 2\pi$

71. To define the inverse tangent, what are the restrictions on the range?

    A. $-\infty < y \leq 0$

    B. $\left[-\dfrac{\pi}{2} < y < \dfrac{\pi}{2}\right]$

    C. $\left[-\dfrac{\pi}{2} \leq y \leq \dfrac{\pi}{2}\right]$

    D. $-2\pi \leq y \leq 2\pi$

72. Find the exact solution(s) for the following equation: $\cos 2\alpha = \cos \alpha$.

    A. $\alpha = 30°, 120°$

    B. $\alpha = 60°, 150°$

    C. $\alpha = 0°, 120°$

    D. $\alpha = 30°, 210°$

73. If the earth has a radius of 4050 miles and makes one rotation every 24 hours, what is the linear velocity of an object situated on the equator?

    A. 880 mph

    B. 940 mph

    C. 1000 mph

    D. 1060 mph

74. Point Q revolves clockwise around a point, making 9 complete rotations in 5 seconds. If Q is 12 cm away from its center of rotation, what is P's angular velocity?

    A. 11.3 radians/sec

    B. 22.6 radians/sec

    C. 135.8 radians/sec

    D. 271.6 radians/sec

75. The horizontal displacement (*d*) of the end of a pendulum is given by the equation
    d = *K*sin2π*t*. What is *K* if *d* = 16 cm and *t* = 4 seconds?

    A. −1260

    B. −260

    C. 260

    D. 1260

Answers

1. A

2. B

3. A

If you missed 1, 2, or 3, go to "Angles and Quadrants," page 21.

4. D

5. C

If you missed 4 or 5, go to "Coterminal Angles," page 25.

6. C

7. A

8. B

If you missed 6, 7, or 8, go to "Trigonometric Functions of Acute Angles," page 28.

9. D

10. C

If you missed 9 or 10, go to "Functions of General Angles," page 38.

11. A

12. C

If you missed 11 or 12, go to "Interpolation," page 45.

13. B

If you missed 13, go to "Understanding Degree Measure," page 59.

14. B

If you missed 14, go to "Understanding Radians," page 59.

15. C

If you missed 15, go to "Relationships between Degrees and Radians," page 60.

16. D

If you missed 16, go to "The Unit Circle and Circular Functions," page 62.

17. A

18. D

If you missed 17 or 18, go to "Domain versus Range," page 63.

19. C

20. B

If you missed 19 or 20, go to "Periodic Functions," page 66.

21. B

If you missed 21, go to "Graphing Sine and Cosine," page 67.

22. C

If you missed 22, go to "Graphing Tangents," page 77.

23. D

If you missed 23, go to "Graphing the Reciprocal Functions," page 78.

24. A

25. D

If you missed 24 or 25, go to "Finding Missing Parts of Right Triangles," page 93.

26. B

27. C

If you missed 26 or 27, go to "Angles of Elevation and Depression," page 95.

28. C

If you missed 28, go to "The Law of Sines," page 100.

29. B

If you missed 29, go to "The Law of Cosines," page 105.

30. B

If you missed 30, go to "SSS," page 109.

31. A

If you missed 31, go to "SAS," page 109.

32. B

If you missed 32, go to "ASA," page 111.

33. A

If you missed 33, go to "SAA," page 111.

34. D

If you missed 34, go to "SSA, The Ambiguous Case," page 111.

35. C

If you missed 35, go to "Areas of Triangles," page 116.

36. C
37. B

If you missed 36 or 37, go to "Heron's Formula (SSS)," page 119.

38. C
39. B

If you missed 38 or 39, go to "Reciprocal Identities," page 137.

40. D
41. A

If you missed 40 or 41, go to "Ratio Identities," page 138.

42. B

If you missed 42, go to "Cofunction Identities," page 139.

43. C

If you missed 43, go to "Pythagorean Identities," page 140.

44. A

If you missed 44, go to "Identities for Negatives," page 139.

45. B
46. C

If you missed 45 or 46, go to "Addition and Subtraction Identities," page 143.

47. A
48. D

If you missed 47 or 48, go to "Double Angle Identities," page 144.

49. A

If you missed 49, go to "Half Angle Identities," page 146.

50. B
51. D

If you missed 50 or 51, go to "Tangent Identities," page 150.

52. B

If you missed 52, go to "Product-Sum and Sum-Product Identities," page 152.

53. C

If you missed 53, go to "Vector versus Scalars," page 167.

54. B

If you missed 54, go to "Vector Addition Triangle/The Tip-Tail Rule," page 169.

55. C
56. C
57. A

If you missed 55, 56, or 57, go to "Parallelogram of Forces," page 172.

58. D

59. D

If you missed 58 or 59, go to "Vectors in the Rectangular Coordinate System," page 178.

60. B

If you missed 60, go to "Algebraic Addition of Vectors," page 183.

61. B

If you missed 61, go to "Scalar Multiplication," page 185.

62. A

If you missed 62, go to "Algebraic Addition of Vectors," page 183.

63. B

If you missed 63, go to "Algebraic Addition of Vectors," page 183.

64. C
65. D
66. A

If you missed 64, 65, or 66, go to "Converting Between Polar and Rectangular Coordinates," page 206.

67. C
68. B

If you missed 67, go to "Plotting Complex Numbers on Rectangular Axes," page 217.
If you missed 68, go to "Plotting Complex Numbers on the Polar Axis," page 219.

69. D

If you missed 69, go to "Finding Powers of Complex Numbers," page 225.

70. C

If you missed 70, go to "Inverse Sine and Cosine," page 244.

71. B

If you missed 71, go to "Inverse Tangent," page 249.

72. C

If you missed 72, go to "Trigonometric Equations," page 257.

73. D

74. A

If you missed 73 or 74, go to "Uniform Circular Motion," page 261.

75. A

If you missed 75, go to "Simple Harmonic Motion," page 263.

Chapter 1
Trigonometric Ideas

The word *trigonometry* comes from two Greek words, *trigonon*, meaning triangle, and *metria*, meaning measurement. This is the branch of mathematics that deals with the ratios between the sides of right triangles with reference to either of its acute angles and enables you to use this information to find unknown sides or angles of any triangle. Trigonometry is not just an intellectual exercise, but has uses in the fields of engineering, surveying, navigation, architecture, and yes, even rocket science.

Angles and Quadrants

An angle is a measure of rotation and is expressed in degrees or radians. For now, we'll stick with degrees, and we'll examine working with radians in the next chapter. Consider any angle in standard position to have its vertex at the origin (the place where the *x*- and *y*-axes cross), labeled *O* in the diagrams. Angle measure is the amount of rotation between the two rays forming the angle.

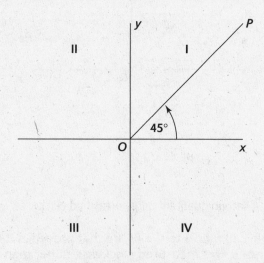

A first quadrant angle in standard position.

Figure the **initial side** of the angle above beginning on the *x*-axis to the right of the origin. Consider the **terminal side** of the angle to be hinged at *O*. The terminal side of the angle, *OP*, was rotated counterclockwise from the *x*-axis through an angle of less than 90° to form the first quadrant angle shown above. Notice the Roman numerals. They mark the quadrants I, or first; II, or second; III, or third; and IV, or fourth. Notice that the quadrant numbers rotate *counterclockwise* around the origin. Because the angle in the above figure has its initial side on the *x*-axis, it is said to be in **standard position.** Had the terminal side made a full turn and come back to the *x*-axis, it would have rotated 360°.

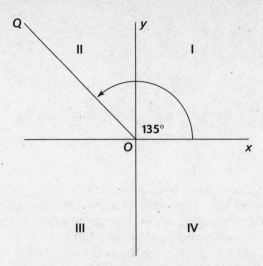

A second quadrant angle in standard position.

The above figure is called a second quadrant angle because its terminal side is in the second quadrant. When the magnitude of an angle is measured in a counterclockwise direction, the angle's measure is positive. The above figure shows an angle of 135° measure.

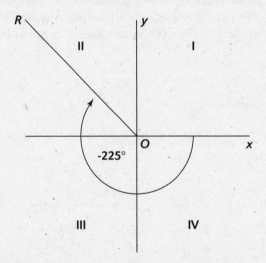

A second quadrant angle measured clockwise.

The angle in the above figure is identical to the figure that precedes it in every way except how the angle was measured. Since it was measured clockwise rather than counterclockwise, it has a measure of −225°. Notice that the absolute value of that angle is obtained by subtracting 135° from 360°. The negative sign marks the direction in which it was measured.

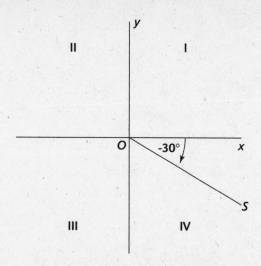

A fourth quadrant negative angle.

Notice that the fourth quadrant angle in the above figure, if measured counterclockwise, would have measured 330°. Can you see why? Moving counterclockwise, it would have been 30° shy of a full 360° rotation.

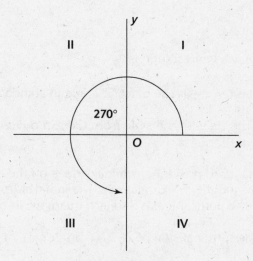

A quadrantal angle.

When an angle is in standard position, and its terminal side coincides with one of the axes, it is referred to as a **quadrantal angle.** Angles of 90°, 180°, and 270° are three examples of quadrantal angles. They are by no means all the quadrantal angles that are possible, but we'll get to that in the next lesson.

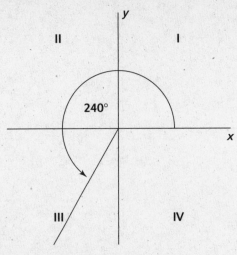

A third quadrant angle.

The one angle remaining to be shown is a Q-III angle (see above). A third quadrant (or Q-III) angle is any angle with its terminal side being in the third quadrant. Because this angle was formed by a counterclockwise rotation, it is positive.

Example Problems

These problems show the answers and solutions.

1. In which quadrant is the terminal side of a 95° angle in standard position?

   answer: II   This also breaks with the style from *CliffsStudySolver Geometry* and *CliffsStudySolver Algebra*.

   Since the angle is in standard position, its initial side is on the *x*-axis to the right of the origin. The *y*-axis forms a right (90°) angle with the initial side, so a 95° angle's terminal side must sweep past the vertical *y*-axis and into quadrant II.

2. In which quadrant is the terminal side of a −320° angle?

   answer: I

   Since the angle is in standard position, its initial side is on the *x*-axis to the right of the origin. Since its sign is negative, its terminal side rotates clockwise past the *y*-axis at −90°, on past the *x*-axis at −180°, past the *y*-axis again at −270°, and continues on another 50° to terminate in the first quadrant. See the figure that follows.

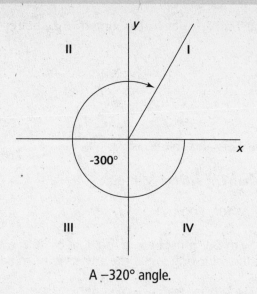

A −320° angle.

3. What is the special name given to a right angle in standard position?

   answer: **quadrantal angle**

   A right angle in standard position will have its terminal side on the y-axis. That makes it a quadrantal angle.

Coterminal Angles

Two angles that are in standard position and share a common terminal side are said to be coterminal angles. All of the angles in the following figure are coterminal with an angle of degree measure of 45°. The arrow shows the direction and the number of rotations through which the terminal side goes.

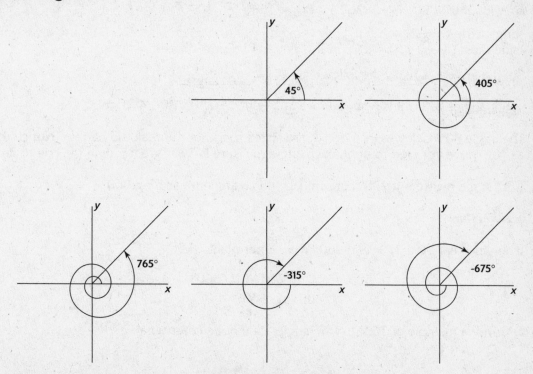

Angles coterminal with 45°.

All angles that are coterminal with an angle measuring $d°$ may be represented by the following equation:

$$d° + n \cdot 360°$$

Example Problems

These problems show the answers and solutions.

1. Name four angles that are coterminal with 80°.

    answer: **–640°, –280°, 440°, 800°, . . .**

    Any angle coterminal with 80° must have a multiple of 360° added to it; according to the formula $d° + n \cdot 360°$, the following are some angles coterminal with 80°:

    $d° + n \cdot 360° = 80° + (1)(360°) = 80° + 360° = 440°$

    $d° + n \cdot 360° = 80° + (2)(360°) = 80° + 720° = 800°$

    $d° + n \cdot 360° = 80° + (3)(360°) = 80° + 1080° = 1160°$

    $d° + n \cdot 360° = 80° + (4)(360°) = 80° + 1440° = 1520°$

    Of course, any number could have been substituted for n, and that means negative as well as positive values, for example:

    $d° + n \cdot 360° = 80° + (-1)(360°) = 80° - 360° = -280°$

    $d° + n \cdot 360° = 80° + (-2)(360°) = 80° - 720° = -640°$

    $d° + n \cdot 360° = 80° + (-3)(360°) = 80° - 1080° = -1000°$

    $d° + n \cdot 360° = 80° + (-4)(360°) = 80° - 1440° = -1360°$

    So, the answer is really the following:

    Four angles coterminal with 80° are . . ., –640°, –280°, 80°, 440°, 800°, . . .

    The angle 80° itself was included in the series so as not to break the pattern. Notice that as you move from left to right, each angle measure is 360° greater than the one to its left.

2. Is an angle measuring 220° coterminal with an angle measuring 960°?

    answer: **No**

    If angles measuring 220° and 960° were coterminal, then

    $$960° = 220° + n \cdot 360°$$
    $$740° = n \cdot 360°$$

But 740° is *not* a multiple of 360°, so the angles *cannot* be coterminal.

Work Problems

Use these problems to give yourself additional practice.

1. In which quadrant does the terminal side of an angle of degree measure 1200° fall?

2. In order for an angle to be a quadrantal angle, by what number must it be capable of being divided?

3. What is the lowest possible positive degree measure for an angle that is coterminal with an angle of −1770°?

4. In which quadrant does the terminal side of an angle of 990° degree measure fall?

5. Name two positive and two negative angles that are coterminal with an angle of degree measure 135°.

Worked Solutions

1. **II** First see how many times 360° can be subtracted or divided out of the total.

   3 · 360° = 1080°

   1200° − 1080° = 120°

   A 120° angle is larger than 90° and less than 180°, and so its terminal side falls in the second quadrant.

2. **90** Quadrantal angles' terminal sides fall on the axes. Therefore, no matter what the size of the coterminal angle, it must be capable of being in the positions of 90°, 180°, 270°, or 0°. All are divisible by 90.

3. **30°** Just keep adding 360° until the sum goes from a negative to a positive value:

   5 · 360° = 1800°

   −1770° + 1800° = 30°

4. **None** It is quadrantal. 990° is two full revolutions from the starting position, and then an additional 270°:

   2 · 360° = 720°

   990° − 720° = 270°

   That places the terminal side on the y-axis, south of the origin (negative).

5. **−585°, −225°, 495°, 855°** Needless to say, there are an infinite number of other solutions, each of which is determined by substituting into the expression: d° + n · 360°.

Trigonometric Functions of Acute Angles

The building blocks of trigonometry are based on the characteristics of similar triangles that were first formulated by Euclid. He discovered that if two triangles have two angles of equal measure, then the triangles are similar. In similar triangles, the ratios of the corresponding sides of one to the other are all equal. Since all right triangles contain a 90° angle, proving two of them similar only requires having one acute angle of one triangle equal in measure to one acute angle of the second. Having established that, we easily find that in two similar right triangles, the ratio of each side to another in one triangle is equal to the ratio between the two corresponding sides of the other triangle. It is no long stretch from there to realize that this must be true of all similar triangles. Those relationships led to the **trigonometric ratios.** It is customary to use lowercase Greek letters to designate the angle measure of specific angles. It doesn't matter which Greek letter is used, but the most common are α (alpha), β (beta), ϕ (phi), and θ (theta).

The trigonometric ratios that follow are based upon the following reference triangle, which is drawn in two different ways.

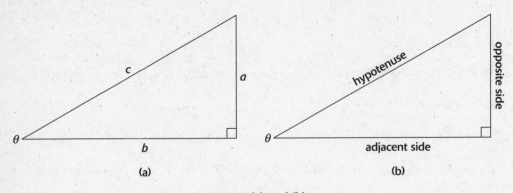

(a)   (b)

Figures (a) and (b).

Both figures show the same triangle with sides a, b, and c, and with angle θ at the left end of the base. The difference is that in figure (b), the two legs are labeled with respect to $\angle\theta$. That is to say, side a is marked as opposite to $\angle\theta$, and side b is adjacent to $\angle\theta$. You might correctly argue that side c is also adjacent to $\angle\theta$, but that side already has a name (you learned about this in plane geometry). Being the side opposite the right angle, it's the hypotenuse, hence it is the nonhypotenuse adjacent side to $\angle\theta$ that is assigned the name "adjacent."

That leads us to the first three trigonometric functions:

The sine of θ is: $\sin\theta = \dfrac{a}{c} = \dfrac{\text{length of side opposite } \theta}{\text{length of hypotenuse}}$

The cosine of θ is: $\cos\theta = \dfrac{b}{c} = \dfrac{\text{length of side adjacent } \theta}{\text{length of hypotenuse}}$

The tangent of θ is: $\tan\theta = \dfrac{a}{b} = \dfrac{\text{length of side opposite } \theta}{\text{length of side adjacent } \theta}$

In the early days of American history, way before the days of political correctness, some ambitious trigonometry student in search of a mnemonic device by which to remember his or her trigonometric ratios dreamed up the SOHCAHTOA Indian tribe, which today would be the SOHCAHTOA tribe of Native Americans. SOHCAHTOA is an acronym for the basic trig ratios and their components; that is:

Sin-**O**pposite/**H**ypotenuse-**C**os-**A**djacent/**H**ypotenuse-**T**an-**O**pposite/**A**djacent

Keep in mind that as long as the angles remain the same, the ratios of their pairs of sides will remain the same, regardless of how big or small they are in length. The trigonometric ratios in right triangles depend exclusively on the angle measurements of the triangles and have no dependence on the lengths of their sides.

Example Problems

These problems show the answers and solutions. All refer to the following figure.

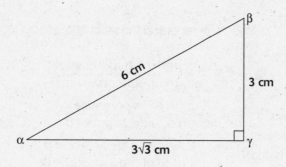

1. Find the sine of α.

    ***answer:* 0.5**

    The sin ratio is $\dfrac{\text{opposite}}{\text{hypotenuse}}$. 3 cm is the length of the side opposite $\angle\alpha$, and the hypotenuse is 6 cm.

    So, $\sin\alpha = \dfrac{3}{6} = \dfrac{1}{2} = 0.5$.

2. Find $\cos\beta$.

    ***answer:* 0.5**

    The cos ratio is $\dfrac{\text{adjacent}}{\text{hypotenuse}}$. 3 cm is the length of the side adjacent $\angle\beta$, and the hypotenuse is 6 cm.

    So, $\cos\beta = \dfrac{3}{6} = \dfrac{1}{2} = 0.5$.

3. Find $\tan\beta$.

    ***answer:* $\sqrt{3}$**

    The tan ratio is $\dfrac{\text{opposite}}{\text{adjacent}}$. $3\sqrt{3}$ cm is the length of the side opposite $\angle\beta$, and its adjacent side is 3 cm. So, $\tan\beta = \dfrac{3\sqrt{3}}{3} = \sqrt{3}$.

Reciprocal Trigonometric Functions

The three remaining trigonometric ratios are the reciprocals of the first three. You may think of them as the first three turned upside down, or what you must multiply the first three by in order to get a product of 1. The reciprocal of the sine is cosecant, abbreviated csc. Secant is the reciprocal of cosine and is abbreviated sec. Finally, cotangent, abbreviated cot, is the reciprocal of tangent.

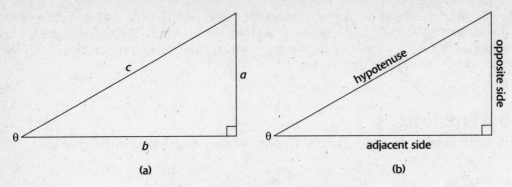

(a)   (b)

Models for reciprocal trigonometric ratios.

The cosecant of θ is: $\csc \theta = \dfrac{c}{a} = \dfrac{\text{length of hypotenuse}}{\text{length of side opposite } \theta}$

The secant of θ is: $\sec \theta = \dfrac{c}{b} = \dfrac{\text{length of hypotenuse}}{\text{length of side adjacent } \theta}$

The cotangent of θ is: $\cot \theta = \dfrac{b}{a} = \dfrac{\text{length of side adjacent } \theta}{\text{length of side opposite } \theta}$

There is no SOHCAHTOA tribe here to help out, but you shouldn't need one. Just remember the pairings, find the right combination for its reciprocal (that is for secant; remember it pairs with cosine), and flip it over.

Example Problems

These problems show the answers and solutions. All problems refer to the following figure.

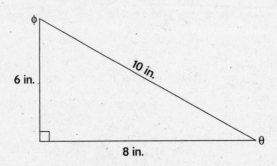

Model for example problems.

1. Find $\csc\phi$.

    answer: **1.25**

    $\csc \phi = \dfrac{\text{length of hypotenuse}}{\text{length of side opposite } \phi} = \dfrac{10}{8} = 1.25$

2. Find $\sec\theta$.

    answer: **1.25**

    $\sec \theta = \dfrac{\text{length of hypotenuse}}{\text{length of side adjacent } \theta} = \dfrac{10}{8} = 1.25$

3. Find $\cot \theta$.

    answer: **1.33**

    $$\cot \theta = \frac{\text{length of side adjacent } \theta}{\text{length of side opposite } \theta} = \frac{8}{6} = 1.33$$

Work Problems

Use these problems to give yourself additional practice.

1. An angle has a sine of 0.3.

    a.   What other function's value can you determine?

    b.   What is that value?

2. An angle has a cosine of 0.6.

    a.   What other function's value can you determine?

    b.   What is that value?

3. An angle has a tangent of 2.5.

    a.   What other function's value can you determine?

    b.   What is that value?

4. An angle has a secant of 1.8.

    a.   What other function's value can you determine?

    b.   What is that value?

5. An angle has a cosecant of 5.3.

    a.   What other function's value can you determine?

    b.   What is that value?

Worked Solutions

1. **cosecant, 3.33**   Sine's reciprocal function is cosecant. It is the reciprocal of sine; that is, $\csc \theta = \frac{1}{\sin \theta}$, but you know that the value of sin is 0.3, therefore $\csc \theta = \frac{1}{0.3}$.

    Divide 1 by 0.3, and you get 3.33 (rounding to the hundredths).

2. **secant, 1.67**   Cosine's reciprocal function is secant. Since it's the reciprocal of cosine, $\sec \theta = \frac{1}{\cos \theta}$, but you were given the cos as 0.6, therefore $\sec \theta = \frac{1}{0.6}$.

    Divide 1 by 0.6 to find a rounded value of 1.67.

3. **cotangent, 0.4**   Tangent's reciprocal function is cotangent. It is the reciprocal of tangent, so $\cot \theta = \frac{1}{\tan \theta}$, but you were given tan = 2.5, so $\cot \theta = \frac{1}{2.5}$.

    Divide 1 by 2.5 and get 0.4.

4. **cosine, 0.56**   Since secant's reciprocal function is cosine, $\cos\theta = \dfrac{1}{\sec\theta}$, but you know
   that the value of sec is 1.8; therefore $\cos\theta = \dfrac{1}{1.8}$. Divide 1 by 1.8, and you get 0.56
   (rounding to the hundredths).

5. **sine, 0.19**   Since cosecant's reciprocal function is sine, $\sin\theta = \dfrac{1}{\csc\theta}$, but you know that
   the value of csc is 5.3; therefore $\sin\theta = \dfrac{1}{5.3}$.

   Divide 1 by 5.3, and you'll get 0.188, which you'll round to 0.19.

Introducing Trigonometric Identities

When trigonometric functions of an angle ϕ are related in an equation, and that equation is true
for all values of ϕ, then the equation is known as a **trigonometric identity.** The following trigonometric identities can be constructed from the trigonometric ratios that you just reviewed.

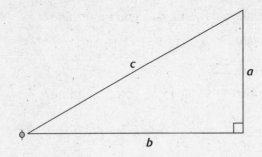

Triangle referenced by identities below.

Referring to the above figure, since $\sin\phi = \dfrac{a}{c}$, $\cos\phi = \dfrac{b}{c}$, and $\tan\phi = \dfrac{a}{b}$, it follows that

$$\tan\phi = \frac{\sin\phi}{\cos\phi} = \frac{\frac{a}{c}}{\frac{b}{c}}.$$

The second part of which you can simplify like this: $\dfrac{a}{c} \cdot \dfrac{c}{b} = \dfrac{a}{\phi} \cdot \dfrac{\phi}{b} = \dfrac{a}{b}$

Which serves to prove the identity: $\tan\phi = \dfrac{\sin\phi}{\cos\phi}$

You could also prove the identity: $\cot\phi = \dfrac{\cos\phi}{\sin\phi}$

These two identities are extremely useful and should be memorized. There's a third very handy
identity, but first, you must become familiar with some conventional notation. The symbol $(\sin\theta)^2$
and $\sin^2\theta$ mean the same thing and may be used interchangeably. With that in mind, the third
identity referred to is $\sin^2\phi + \cos^2\phi = 1$. If you would like to see that proven, refer to the previous figure and the Pythagorean theorem, stated as $a^2 + b^2 = c^2$.

$$\sin^2\phi + \cos^2\phi = \left(\frac{a}{c}\right)^2 + \left(\frac{b}{c}\right)^2 = \frac{a^2}{c^2} + \frac{b^2}{c^2}$$

You can add the two fractions to get $\dfrac{a^2 + b^2}{c^2}$.

But you already know that $a^2 + b^2 = c^2$, so you substitute in the numerator and simplify:

$$\frac{c^2}{c^2} = 1$$

Therefore, $\sin^2\phi + \cos^2\phi = 1$.

The importance of these three identities cannot be over-stressed. You will deal much more extensively with identities in the fourth chapter, but for now, try to learn these three.

Example Problems

These problems show the answers and solutions. Where applicable, round each answer to the nearest thousandth.

1. Find the sin and tangent of $\angle\lambda$ if λ is an acute angle $(0° < \lambda < 90°)$ and $\cos\lambda = \cos\lambda = \frac{1}{5}$.

   answer: **0.980, 4.9**

   Since $\sin^2\phi + \cos^2\phi = 1$, then $\qquad\qquad\qquad \sin^2\lambda + \cos^2\lambda = 1$.

   Substitute: $\qquad\qquad\qquad\qquad\qquad\qquad \sin^2\lambda + \left(\frac{1}{5}\right)^2 = 1$

   Square and subtract from both sides: $\qquad \sin^2\lambda = 1 - \frac{1}{25}$

   Subtract: $\qquad\qquad\qquad\qquad\qquad\qquad\quad \sin^2\lambda = \frac{24}{25}$

   Take the square root of both sides: $\qquad\quad \sin\lambda = \sqrt{\frac{24}{25}} = 0.980$

   Next, find the tangent using what you just found: $\quad \cos\lambda = \tan\lambda = \dfrac{\sin\lambda}{\cos\lambda}$ and

   $$\cos\lambda = \frac{1}{5} = 0.2.$$

   Substituting, you find that $\qquad\qquad\qquad \tan\lambda = \dfrac{0.980}{0.2} = 4.9$.

2. Find the cos and tangent of $\angle\phi$ if ϕ is an acute angle $(0 < \phi < 90°)$ and $\sin\phi = 0.867$.

   answer: **0.498, 1.741**

   First, use the identity $\sin^2\phi + \cos^2\phi = 1$ to find $\cos\phi$.

   Substitute: $\qquad\qquad (0.867)^2 + \cos^2\phi = 1$

   Subtract $(0.867)^2$ from both sides: $\quad \cos^2\phi = 1 - (0.867)^2$

   Square the quantity in parentheses: $\quad \cos^2\phi = 1 - 0.752$

   Subtract 0.752 from 1: $\qquad\qquad\quad \cos^2\phi = 0.248$

   Take the square root of both sides: $\quad \cos\phi = 0.498$

   Now, solve for the tangent.

First write the relevant identity: $\tan\phi = \dfrac{\sin\phi}{\cos\phi}$

Substitute for sin and cos: $\tan\phi = \dfrac{0.867}{0.498}$

And divide: $\tan\phi = 0.741$

3.   Find the sin of $\angle\theta$ if θ is an acute angle ($0 < \theta < 90°$), $\tan\theta = 1.192$, and $\cos\phi = 0.643$.

answer: 0.766

This time, you only need: $\tan\theta = \dfrac{\sin\theta}{\cos\theta}$

Substitute what you know: $1.192 = \dfrac{\sin\theta}{0.643}$

Now multiply both sides by 0.643: $(0.643)(1.192) = \left(\dfrac{\sin\theta}{0.643}\right)(0.643)$

Reversing the results, you get: $\sin\theta = 0.766.$

Trigonometric Cofunctions

Trigonometric functions are often considered in pairs, known as cofunctions. Sine and cosine are cofunctions. So are secant and cosecant. The final pair of cofunctions are tangent and cotangent. From the right triangle *ABC*, the following identities can be seen:

$$\sin A = \frac{a}{c} = \cos B \qquad\qquad \sin B = \frac{b}{c} = \cos A$$

$$\sec A = \frac{c}{b} = \csc B \qquad\qquad \sec B = \frac{c}{a} = \csc A$$

$$\tan A = \frac{a}{b} = \cot B \qquad\qquad \tan B = \frac{b}{a} = \cot A$$

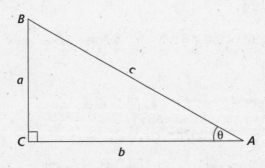

Reference triangle for cofunctions.

To refresh your memory, all three angles of a triangle are supplementary (sum to 180°), and angles that sum to 90° are known as complementary. Since one of the three angles in a right triangle measures 90°, the sum of the remaining acute angles must be complementary. Refer to the above reference triangle to confirm the following relationships:

$\sin\theta = \cos(90° - \theta)$ $\cos\theta = \sin(90° - \theta)$
$\sec\theta = \csc(90° - \theta)$ $\csc\theta = \sec(90° - \theta)$
$\tan\theta = \cot(90° - \theta)$ $\cot\theta = \tan(90° - \theta)$

Two Special Triangles

The figure below shows an isosceles right triangle with each leg having a length of 1. Can you figure out the measure of angle X?

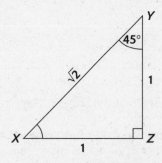

An isosceles right triangle.

That last question was intended as a joke, since in any isosceles triangle, the angles opposite the equal legs are always equal in measure. What is different, however, is that in an isosceles right triangle, the acute angles always measure 45°. That's because they must be both complementary and equal. What is also different in an isosceles right triangle is that the hypotenuse always has the same relationship to the legs. That relationship, of course, may be found using the Pythagorean theorem.

In this case: $z^2 = x^2 + y^2$

Substituting: $z^2 = 1^2 + 1^2$

Squaring and adding: $z^2 = 2$

Therefore: $z = \sqrt{2}$

All right, you already knew that, because it shows that in the above figure, but what if instead of 1, x and y had been 2?

Substituting: $z^2 = 2^2 + 2^2$

Squaring and adding: $z^2 = 8$

Therefore: $z = \sqrt{8} = \sqrt{4 \cdot 2} = 2\sqrt{2}$

Do you see the pattern yet? Try one more, just to make sure. This time, let x and y be 5.

Substituting: $z^2 = 5^2 + 5^2$

Squaring and adding: $z^2 = 50$

Therefore: $z = \sqrt{50} = \sqrt{25 \cdot 2} = 5\sqrt{2}$

To sum it all up, when dealing with an isosceles right triangle, the hypotenuse is always the length of the leg times the square root of 2.

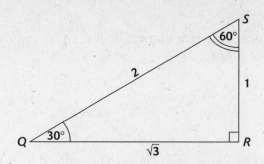

The 30-60-90 right triangle.

Students and teachers of trigonometry are quite fond of a second special right triangle. That's the one with acute angles of 30° and 60°, or as it's often referred to, the 30-60-90 right triangle. The relationship among the sides are spelled out in the above figure. Notice that the side opposite the 30° angle is half the length of the hypotenuse, and the length of the side opposite the 60° angle is half the hypotenuse times the square root of three.

From the two special triangles, you can compile a table of frequently used trigonometric functions.

Table of Trigonometric Ratios for 30°, 45°, and 60° Angles						
θ	$\sin\theta$	$\cos\theta$	$\sec\theta$	$\csc\theta$	$\tan\theta$	$\cot\theta$
30°	$\dfrac{1}{2}$	$\dfrac{\sqrt{3}}{2}$	$\dfrac{2\sqrt{3}}{3}$	2	$\dfrac{\sqrt{3}}{3}$	$\sqrt{3}$
45°	$\dfrac{\sqrt{2}}{2}$	$\dfrac{\sqrt{2}}{2}$	$\sqrt{2}$	$\sqrt{2}$	1	1
60°	$\dfrac{\sqrt{3}}{2}$	$\dfrac{1}{2}$	2	$\dfrac{2\sqrt{3}}{3}$	$\sqrt{3}$	$\dfrac{\sqrt{3}}{3}$

Example Problems

These problems show the answers and solutions.

1. One leg of an isosceles right triangle is 4 cm long.

    a. How long is the other leg?

    b. How long is the hypotenuse?

    answer: **4 cm, $4\sqrt{2}$**

    For part a, both legs of an isosceles right triangle are the same length. As for part b, you have seen that the hypotenuse of an isosceles right triangle is equal in length to a side times the square root of two.

2. The shortest leg of a 30-60-90 triangle is 5 inches long.

    a. How long is the other leg?

    b. How long is the hypotenuse?

    answer: **$5\sqrt{3}$, 10 in.**

It might be simpler to find the answer to part b first. In a 30-60-90 triangle, the shortest side is the one opposite the smallest angle, that is, the one opposite 30°. That side is half the length of the hypotenuse, so the hypotenuse must be twice the length of that side, or 10 in. Finally, the side opposite the 60° angle (the other leg) is half the hypotenuse times the square root of three.

3. In triangle ABC, with right angle at B, the cosine of $\angle A$ is $\frac{11}{15}$. What is the sine $\angle C$?

    answer: $\frac{11}{15}$

    The cofunction identities tell us that in a given right triangle, the sine of one acute angle is the cosine of the other.

Work Problems
Use these problems to give yourself additional practice.

1. In triangle ABC, with right angle at B, the cosine of $\angle A$ is $\frac{3}{5}$. What are the sine of $\angle A$ and the tangent of $\angle A$?

2. One leg of an isosceles right triangle is 8 cm long. What is the length of the hypotenuse?

3. The hypotenuse of a 30-60-90 triangle is 10 inches long. The shorter leg of the triangle is a, and the longer b. Find the lengths of a and b.

4. In triangle PQR, with right angle at P, the sine of $\angle Q$ is 0.235. What is the tangent of $\angle R$?

Worked Solutions

1. $\frac{4}{5}, \frac{4}{3}$

Remember:	$\sin^2 A + \cos^2 A = 1$
Therefore:	$\sin^2 A + \left(\frac{3}{5}\right)^2 = 1$
Square and subtract from both sides:	$\sin^2 A = 1 - \frac{9}{25}$
Make 1 an equivalent fraction:	$\sin^2 A = \frac{25}{25} - \frac{9}{25}$
Subtract:	$\sin^2 A = \frac{16}{25}$
Now get the square root of both sides:	$\sin A = \frac{4}{5}$
As for tangent, use the ratio identity:	$\tan A = \frac{\sin A}{\cos A}$
Substitute:	$\tan A = \frac{\frac{4}{5}}{\frac{3}{5}}$
Rewrite as a reciprocal multiplication:	$\tan A = \frac{4}{5} \times \frac{5}{3}$
The 5s cancel, so you get:	$\tan A = \frac{4}{\cancel{5}} \times \frac{\cancel{5}}{3} = \frac{4}{3}$

2. **$8\sqrt{2}$ cm**   You learned in the "Two Special Triangles" section that the hypotenuse of an isosceles right triangle is equal to the length of a side times the square root of two. If you did not recall that, then use the Pythagorean theorem, which in the case of an isosceles right triangle may be written

$c^2 = a^2 + a^2$ (Remember, both legs are equal.)

Substitute: $c^2 = 8^2 + 8^2$

Square and add: $c^2 = 64 + 64 = 128$

Solve for c: $c = 8\sqrt{2}$

3. **5 inches, $5\sqrt{3}$ inches**   In a 30-60-90 triangle, the shorter leg is opposite the 30° angle and is half the length of the hypotenuse. Half of 10 inches is 5 inches. The longer leg is opposite the 60° angle and is equal to half the hypotenuse times the square root of 3. That's $5\sqrt{3}$ inches.

4. **17.67**   The sine of $\angle Q$ is 0.235, but you want the tangent of $\angle R$, so you'll use sine's cofunction, $\cos\angle R = 0.235$. Next, you need to find $\sin\angle R$ so that you may relate sin and cos with the tangent identity.

First write the equation: $\sin^2 R + \cos^2 R = 1$

Next, substitute: $\sin^2 R + (0.235)^2 = 1$

Clear the parentheses: $\sin^2 R + 0.055 = 1$

Collect the constants: $\sin^2 R = 1 - 0.055$

Subtract: $\sin^2 R = 0.945$

Solve for $\sin R$: $\sin R = 0.972$

Now for the tangent identity: $\tan R = \dfrac{\sin R}{\cos R}$

Substitute: $\tan R = \dfrac{0.972}{0.055}$

And divide: $\tan R = 17.67$

Functions of General Angles

When an acute angle is written in standard position, it is always in the first quadrant, and there, all trigonometric functions exist and are positive. This is not true, however, of angles in general. Some of the six trigonometric functions are undefined for quadrantal angles, and some have negative values in certain quadrants. In standard position, an angle is considered to have its starting position in quadrant I on the x-axis and its terminal side in or between one of the four quadrants.

Consider the angle, ϕ in the following figure. Point P is on the terminal side of the angle, r, and has coordinates (x, y). The radius of the circle is 1.

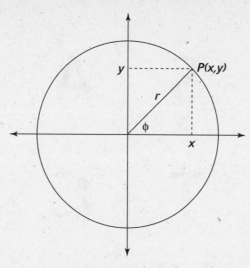

A unit circle ($r = 1$).

Look it over carefully, and you'll see that $\sin\phi = y$, $\cos\phi = x$, and $\tan\phi = \frac{y}{x}$.

This is a tangible proof of the $\tan\phi = \dfrac{\sin\phi}{\cos\phi}$ ratio identity.

Alas, not all standard angles terminate in the first quadrant. Look at these three angles:

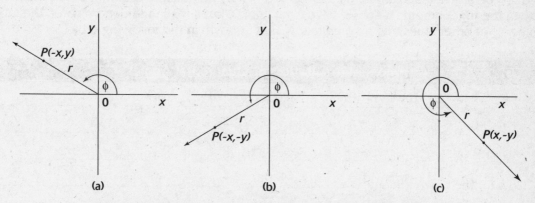

Angles terminating in quadrants other than I: (a) Angle terminating in II;
(b) Angle terminating in III; (c) Angle terminating in IV

In all four quadrants, the value of r is positive. In quadrant II, only sine is positive $\left(\dfrac{y}{r}\right)$, while cosine $\left(\dfrac{-x}{r}\right)$ and tangent $\left(\dfrac{y}{-x}\right)$ are negative. Cosecant, secant, and tangent will always have the same signs as their inverses. We leave it to you to figure out what functions are positive or negative in quadrants III and IV, and why.

To help in the future, there's a little mnemonic scheme you might want to remember, and it's represented in the following figure.

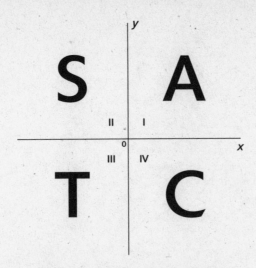

Clockwise ACTS, or counter CAST.

You may remember this as being the word "CAST" reading counterclockwise and beginning in quadrant IV, or "ACTS" reading clockwise from quadrant I. The letters tell you which ratio (and its inverse) is positive in that quadrant: "C" for cosine in IV, "T" for tangent in III, "S" for sine in II, and "A" for all in I.

Should $\angle\phi$ be a quadrantal angle, then either x or y may be equal to 0. If that 0 is in the numerator, then the trigonometric ratio will have a value of 0, but if it's in the denominator, the ratio is undefined. Yet other times, the ratio equals 1. That's shown in the following table.

Trigonometric Ratios for Quadrantal Angles						
θ	$\sin\theta$	$\cos\theta$	$\tan\theta$	$\sec\theta$	$\csc\theta$	$\cot\theta$
0°	0	1	0	1	—	—
90°	1	0	—	—	1	0
180°	0	−1	0	−1	—	—
270°	−1	0	—	—	−1	0

Example Problems

These problems show the answers and solutions.

1. What is the sign of tan 230°?

   answer: **positive**

   Refer to the previous figure. 230° falls in quadrant III, where tangent is positive.

2. What is the sign of sin 300°?

   answer: **negative**

   Refer to the previous figure. 300° falls in quadrant IV, where sine is negative.

3. What is the sign of sec 320°?

    ***answer:* positive**

    Secant is the reciprocal of cosine. 320° falls in quadrant IV, where cosine is positive, so secant must also be positive in that quadrant.

Reference Angles

The trigonometric functions of nonacute angles may be converted so that they correspond to the functions of acute angles. See the following figure.

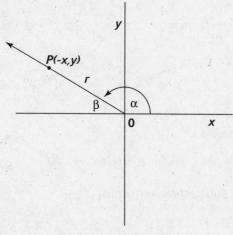

$$180° - \alpha = \text{reference}\angle\beta.$$

α is a second quadrant angle in standard position. By subtracting $\angle\alpha$ from 180°, you get the acute reference angle, $\angle\beta$. You can now find the trigonometric ratios using the reference angle, but bear in mind that since the original angle terminated in the second quadrant, all of the ratios will be negative except for sine and cosecant.

$$\sin\beta = \frac{y}{r} \qquad \cos\beta = \frac{-x}{r} \qquad \tan\beta = \frac{y}{-x}$$

To find cosecant, secant, and cotangent, just flip over the three preceding ratios.

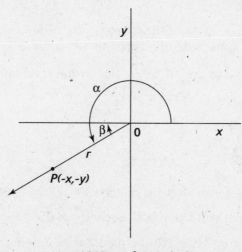

$$\alpha - 180° = \text{reference}\angle\beta.$$

In the third quadrant, reference $\angle\beta$ is found by subtracting 180° from α, as shown in the previous figure.

$$\sin\beta = \frac{-y}{r} \qquad \cos\beta = \frac{-x}{r} \qquad \tan\beta = \frac{-y}{-x}$$

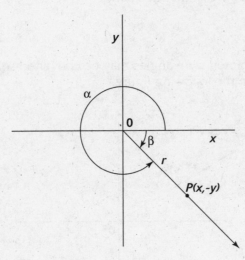

$$360° - \alpha = \text{reference}\angle\beta.$$

In quadrant IV, $\angle\beta$ is found by subtracting α from 360°.

$$\sin\beta = \frac{-y}{r} \quad \cos\beta = \frac{x}{r} \quad \tan\beta = \frac{-y}{x}$$

Notice that the reference angle always sits above or below the *x*-axis.

Example Problems

These problems show the answers and solutions.

1. What is the cosine of 150°?

    answer: **–0.867 or** $-\frac{\sqrt{3}}{2}$

    To find the reference angle, subtract 150° from 180°. That's a 30° angle. In a 30-60-90 triangle of hypotenuse 2, the side opposite the 30° angle is 1, and the side opposite the 60° angle is $\sqrt{3}$. That makes $\cos 30° = \frac{\sqrt{3}}{2}$, but cosine is negative in the second quadrant. $\sqrt{3} \approx 1.732$, hence the two possible answers.

2. What is the sine of 225°?

    answer: **–0.707 or** $-\frac{\sqrt{2}}{2}$

    To find a quadrant III reference angle, subtract 180°. 225° – 180° = 45°. When you studied an isosceles right triangle, you found both sin and $\cos 45° = \frac{1}{\sqrt{2}}$, which you rationalize by multiplying it by $\dfrac{\sqrt{2}}{\sqrt{2}}$, thus getting $\dfrac{\sqrt{2}}{2}$. It's considered poor form to leave a radical in the denominator. $\sqrt{2} \approx 1.414$, so half of it is 0.707, but since sine is negative in the third quadrant, the solution is negative.

3. Find the tangent of 300°.

   answer: −1.732 or $-\sqrt{3}$

   The angle is in quadrant IV, so subtract it from 360°. 360° − 300° = 60°. Refer back to solution 1, to find that the $\tan 60° = \dfrac{\sqrt{3}}{1} = \sqrt{3}$. In quadrant IV, tan is negative, hence the answers shown.

Squiggly versus Straight

We presume that you noticed the squiggly equal signs used in the explanations of example problems 1 and 2. Mathematicians use that symbol to indicate that one quantity is a close approximation of another.

For example, $\sqrt{3}$ is easily shown on a calculator to equal 1.732050808, and then some, so mathematicians would much prefer to write $\sqrt{3} \approx 1.732$ than $\sqrt{3} = 1.732$.

On the other hand, a mathematician would never write 5 + 7 ≈ 12. In fact, 5 + 7 = 12, with no approximating or rounding involved. The squiggly equal sign is also likely to be used when the numbers are readings from scientific instruments, which only approximate true conditions.

Work Problems

Use these problems to give yourself additional practice.

1. A 135° angle is angle in standard position. What is its cosecant?

2. Find the cotangent of a 240° angle in standard position.

3. What is the sine of a 675° angle in standard position?

4. Find the secant of a 330° angle in standard position.

5. Angle χ is in standard position, and its terminal side passes through the point with coordinates (−12,5). Find all six of its trigonometric functions.

6. If $\sin\theta = \dfrac{4}{5}$, and $\cos\theta$ is negative, what are the values of the five remaining trigonometric functions?

Worked Solutions

1. **$\sqrt{2}$ or 1.414**　The reference angle for 135° is 45°. If necessary, refer back to the figure on p. 35 ("An isosceles right triangle"), where you'll find that csc 45° is $\frac{\sqrt{2}}{1}$, which, of course, is simply $\sqrt{2}$.

   Since sine is positive in the second quadrant, so is cosecant.

2. **$\frac{\sqrt{3}}{3}$ or 0.577**　To find the reference angle, subtract 180°. That makes 60°. Next, if necessary, look back at the figure on p. 36 ("The 30-60-90 right triangle").

   $\cot 60° = \frac{1}{\sqrt{3}}$, which you rationalize: $\frac{1}{\sqrt{3}} \cdot \frac{\sqrt{3}}{\sqrt{3}} = \frac{\sqrt{3}}{3}$

   Finally, consider that the angle originally terminated in Q-III (third quadrant), where tangent, and therefore, cotangent, is positive.

3. **$-\frac{\sqrt{2}}{2}$ or −.707**　675° goes one full 360° turn, and then 45° less than another one. That makes it a 45° angle in Q-IV, just below the x-axis. sin 45°, as you may recall from the figure on p. 35 ("An isosceles right triangle") or the table on p. 36 (Table of Trigonometric Ratios for 30°, 45°, and 60° Angles), $= \frac{\sqrt{2}}{2}$. Sine is not positive in Q-IV, so the answer will be negative.

4. **$\frac{2\sqrt{3}}{3}$ or 1.155**　Subtract from 360°, and you find that this is a 30° angle in Q-IV.

   Secant 30° is $\frac{2\sqrt{3}}{3}$. Since cosine is positive in Q-IV, so is secant.

5. Since the angle's terminal side passes through (−12, 5), build a right triangle about that point in Q-II, with the angle to be dealt with at the origin. The side opposite that angle is the y-coordinate, 5. The adjacent side of the triangle is the absolute value of the x-coordinate, $|-12| = 12$. To find the hypotenuse of the triangle, you can either use the Pythagorean theorem, or remember the Pythagorean triple, 5-12-13. Either way, the hypotenuse has a length of 13. So, take care to remember that the terminal side of the angle is in Q-II, where only sine and its reciprocal, cosecant, are positive. With that in mind, you find:

   $\sin\chi = \frac{5}{13}$, $\cos\chi = -\frac{12}{13}$, $\tan\chi = -\frac{5}{12}$, $\csc\chi = \frac{13}{5}$, $\sec\chi = -\frac{13}{12}$, and $\cot\chi = -\frac{12}{5}$

6. **Since $\sin\theta = \frac{4}{5}$ and $\cos\theta$ is negative, the angle must be in the second quadrant; since in Q-I, cosine would have been positive, and sine is positive in Q-III and Q-IV.** The hypotenuse of the triangle must be 5, and the opposite side 4 (from the sine), which makes the adjacent side 3, either by the Pythagorean theorem, or by remembering the Pythagorean triple, 3-4-5. So, $\cos\theta = -\frac{3}{5}$ and $\tan\theta = -\frac{4}{3}$, $\csc\theta = \frac{5}{4}$, $\sec\theta = -\frac{5}{3}$, and $\cot\theta = -\frac{3}{4}$.

Trig Tables versus Calculators

We've already taken note of the fact that the trigonometric ratios depend only on the value of the angle measure and are independent of the size of the triangle's sides. You would think that somebody would write down those values for each degree of measure. It should come as no surprise then, that somebody has. You'll find a table of trigonometric ratios in the back of this book, beginning on p. 297. In order to read the value of a trigonometric ratio for a certain angle, you move one finger down the column that says the name of the ratio you're looking for at the top, while moving your eyes down the angle values on the left until you find the one for which you're looking. Then move them across to the proper column. Where your eyes and finger meet should be the value you want. The tables show values for all six trigonometric functions in increments of 1°.

Now all of the angles we've spoken about so far have been in degrees, but the degree is not the smallest unit of angle measure. Just like that other round thing we're used to seeing on a wall, degrees can be broken down into minutes. There are, in fact, 60 minutes in one degree, and, in turn, 60 seconds in one minute. What that means is that it is possible to break the degree measurement of a circle into $360 \times 60 \times 60 = 1{,}296{,}000$ parts. All of those parts are not especially useful or accurate when dealing with the small circles in this book or on your graph paper, but consider how handy they can be when describing the location on a three-dimensional circular object, like the earth.

In addition to the traditional subdivisions mentioned in the last paragraph, degrees are also capable of being subdivided into decimal parts, like 33.5°, 40.75°, or even 27.625°. There is really no limit to the number of decimal places to which an angle's size may be carried out, although after the sixth, one might start to become a bit suspicious about how accurately that quantity was measured. It is also true, however, that scientific measuring techniques are continually being refined.

That brings us to scientific calculators. Scientific calculators have the capacity to calculate at least the three fundamental trigonometric functions: sine, cosine, and tangent. The method in which the data is entered varies from brand to brand, but for the most part, you press the button that names the ratio you want to find, enter the number of degrees and any decimal portion that may follow, and press <Enter> to display the ratio. Most calculators do not provide cosecant, secant, or cotangent keys, however. In order to find the cosecant of, say 30°, you would first press the <SIN> button, type 30, and press <Enter>. The screen now displays 0.5. Next, clear the screen or add the answer to memory and clear the screen. Next type 1÷, or 1/ depending upon how your calculator is labeled; they mean the same thing. Then either press your <RECALL> key or key in .5. Finally, press <Enter>. The screen will show 2, the cosecant of 30°. That's because, as you'll hopefully recall, cosecant and sine are inverse functions:

$$\left(\csc \phi = \frac{1}{\sin \phi} \right)$$

When using a table of trigonometric functions, it's possible to look at the function and see what the angle is. Go ahead. Look at the sines in the table on p. 297 (Trig Functions Table) and find the angle whose sine is 0.423. Go ahead, really do it. We'll wait. Well, if you actually went and looked under the sin column and found the value 0.423, you would have found that it belongs to the angle, 25°. You can't do the same thing with a calculator. Instead, you have to use the functions arcsin, arccos, and arctan, which are also written $\sin^{-1}$, $\cos^{-1}$, and $\tan^{-1}$. That means, the expression $\sin \theta = 0.5$ may be written as $\sin^{-1} \theta = 0.5$ or $\arcsin \theta = 0.5$. Sometimes, it is referred to as "the inverse sine of 0.5." In any of those cases, the expression is used if the trigonometric ratio is known, and the angle that it belongs to is being sought. In the last case, when $\sin^{-1} \theta = 0.5$, then $\theta = 30°$.

Calculators handle arcsine in different ways. On some, you push an <arc> button and then press the appropriate trig ratio. On others, you must press the <2nd function> button and then press the appropriate trig ratio button.

Interpolation

Most trigonometric functions, whether listed in tables or found on calculators, are approximations. Nevertheless, every effort possible is made to get the numbers as close as possible to the actual value. Bearing that in mind, if you're using tables that give values to the nearest degree, and you need the cosine of a fractional angle, say 30.6°, you're going to need a way to find a value that is more accurate than 0.8660, the cosine for 30°. The method for approximating the closer value is known as **interpolation.** It works like this:

$$\cos 30° > \cos 30.6° > \cos 31° \text{ (which means } \cos 30.6° \text{ is between } \cos 30° \text{ and } \cos 31°)$$
$$\cos 30° = 0.8660, \cos 31° = 0.8572.$$
$$\text{The difference between them is } 0.8660 - 0.8572 = 0.0088.$$

30.6° is 0.6, or $\frac{6}{10}$ of the way between the two angles, so cos 30.6° should be 0.6 of the difference between the two cosines less than cos 30°: $0.0088 \times 0.6 = 0.0053$.

Since cos 31° is lower than cos 30°, subtract from cos 30°:

$$0.8660 - 0.0053 = 0.8607$$

$$\therefore \cos 30.6° = 0.8607 \text{ (The symbol } \therefore \text{ means "therefore.")}$$

Let's try one more of those, so that you have a chance to see both possible scenarios. This time, find the sine of 42.3°.

$$42° < 42.3° < 43°$$
$$\sin 42° = 0.6691, \sin 43° = 0.6820.$$
$$\text{The difference between them is } 0.6820 - 0.6691 = 0.0129.$$

42.3° is 0.3, or $\frac{3}{10}$ of the way between the two angles, so sin 42.3° should be 0.3 of the difference between the two sines greater than sin 42°: $0.0129 \times 0.3 = 0.0039$.

Since sin 42° is lower than sin 43° add to sin 42°:

$$0.6691 + .0039 = 0.6730$$
$$\therefore \sin 42.3° = 0.6730$$

Interpolation may also be used to approximate the size of an angle to the nearest tenth of a degree. Suppose that you have $\sin\theta = 0.9690$, and you know that sin 75° = 0.9659 and sin 76° = 0.9702.

$$1.0 \left\{ x \left\{ \begin{array}{l} \sin 75° = 0.9659 \\ \sin \phi = 0.9690 \\ \sin 76° = 0.9702 \end{array} \right\} 0.0031 \right\} 0.0043 \, (x = \text{the difference})$$

Now, we'll use the variable x to set up a proportion:

$$\frac{x}{1.0} = \frac{0.0031}{0.0043}$$
$$0.0043x = 0.0031$$
$$x = \frac{0.0031}{0.0043}$$
$$x = 0.7$$
$$\therefore \phi = 75° + 0.7° = 75.7°$$

Example Problems

These problems show the answers and solutions. Use the tables that begin on p. 297. Do *not* use a calculator.

1. How many seconds are there in a 5° angle?

    answer: **18,000**

    There are 60 minutes in one degree, and 60 seconds in each of those minutes.

    That's for one degree: $60 \times 60 = 3600$

    For 5 degrees, multiply it by 5: $5 \times 3600 = 18,000$

2. Find tan 46.8° to the nearest ten thousandth (four decimal places).

    answer: **1.0650**

    $46° < 46.8° < 47°$

    tan 46° = 1.0355, tan 47° = 1.0724

    The difference between them is $1.0724 - 1.0355 = 0.0369$

    46.8° is 0.8, or $\frac{8}{10}$ of the way between the two angles, so tan 46.8° should be 0.8 of the difference between the two tans greater than tan 46°: $0.0369 \times 0.8 = 0.0295$.

    Since tan 46° is lower than tan 47° add to tan 46°:

    $$1.0355 + .0295 = 1.0650$$
    $$\therefore \tan 46.8° = 1.0650$$

3. For what acute angle does cosine have a value of 0.7826 if cos 38° = 0.7880 and cos 39° = 0.7771?

    answer: **38.5°**

    $$1.0 \left\{ x \left\{ \begin{array}{l} \cos 75° = 0.7880 \\ \cos \phi = 0.7826 \\ \cos 76° = 0.7771 \end{array} \right\} 0.0054 \right\} 0.0109 \, (x = \text{the difference})$$

    Now, we'll use the variable x to set up a proportion:

    $$\frac{x}{1.0} = \frac{0.0054}{0.0109}$$

    $$0.0109x = 0.0054$$

    $$x = \frac{0.0054}{0.0109}$$

    $$x = 0.5$$

    $$\therefore \phi = 38° + 0.5° = 38.5°$$

Work Problems

Use these problems to give yourself additional practice.

1. How many seconds are there in a 30° angle?

2. Find sin 35.7° to the nearest ten thousandth (four decimal places).

3. Find tan 61.3° to the nearest ten thousandth (four decimal places).

4. What is the secant of a 765° angle in standard position?

5. For what acute angle does sine have a value of 0.4602 if sin 27° = 0.4540 and sin 28° = 0.4695?

Worked Solutions

1. **108,000**   60 seconds per minute × 60 minutes per hour × 30 degrees gives

$$60 \times 60 \times 30 = 108,000$$

2. **0.5835**   35° < 35.7° < 36°

sin 35° = 0.5736, sin 36° = 0.5878.

The difference between them is 0.5878 − 0.5736 = 0.0142.

35.7° is 0.7, or $\frac{7}{10}$ of the way between the two angles, so sin 35.7° should be 0.7 of the difference between the two sines greater than sin 35°: 0.0142 × 0.7 = 0.0099.

Since sin 35° is lower than sin 36° add to sin 35°:

$$0.5736 + .0099 = 0.5835$$
$$\therefore \sin 35.7° = 0.5835$$

3. **1.8270**   61° < 61.3° < 62°

tan 61° = 1.8040, tan 62° = 1.8807.

The difference between them is 1.8807 − 1.8040 = 0.0767.

61.3° is 0.3, or $\frac{3}{10}$ of the way between the two angles, so sin 61.3° should be 0.3 of the difference between the two tans greater than tan 61°: 0.0767 × 0.3 = 0.0230.

Since tan 61° is lower than tan 62° add to tan 61°:

$$1.8040 + .0230 = 1.827$$
$$\therefore \tan 61.3° = 1.827$$

4.　**1.4142**　First of all, a 765° angle is a 45° angle after the two full rotations are removed. Secant is the inverse of cosine, so find cos 45° = .7071

Then make the fraction $\dfrac{1}{0.7071}$ and divide 1 ÷ 0.7071 = 1.4142.

Alternately, you could have recalled that $\cos 45° = \dfrac{\sqrt{2}}{2}$, so $\sec 45° = \dfrac{2}{\sqrt{2}}$.

You can rationalize the denominator by multiplying to get $\dfrac{2}{\sqrt{2}} \cdot \dfrac{\sqrt{2}}{\sqrt{2}} = \dfrac{2\sqrt{2}}{2} = \sqrt{2}$, or 1.4142

5.　**27.4°**　$1.0 \left\{ x \left\{ \begin{matrix} \sin 27° = 0.4540 \\ \sin\theta = 0.4602 \\ \sin 28° = 0.4695 \end{matrix} \right\} 0.0062 \right\} 0.0155\,(x = \text{the difference})$

Now, you'll use the variable x to set up a proportion:

$$\frac{x}{1.0} = \frac{0.0062}{0.0155}$$

$$0.0155x = 0.0062$$

$$x = \frac{0.0062}{0.0155}$$

$$x = 0.4$$

$$\therefore \theta = 27° + 0.4° = 27.4°$$

Chapter Problems and Solutions

Problems

Solve these problems for more practice applying the skills from this chapter. Worked out solutions follow the problems.

1.　In which quadrant is the terminal side of a 105° angle in standard position?

2.　What is the special name given to a straight angle in standard position?

3.　Is an angle measuring 245° coterminal with an angle measuring 975°?

4.　In order for an angle to be a quadrantal angle, by what number must it be capable of being divided?

5.　In which quadrant does the terminal side of an angle of 1260° degree measure fall?

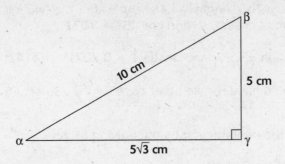

Problems 6–7 refer to the above figure.

6. Find the sin of α.

7. Find tanβ.

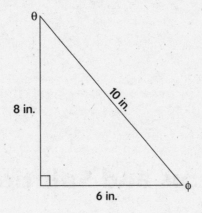

Problem 8 refers to the above figure.

8. Find secθ.

9. An angle has a sine of 0.4. What is the value of its cosecant?

10. An angle has a tangent of 2.5. What is the value of its cotangent?

11. An angle has a cosecant of 3.8. What is the value of its sine?

12. Find the cos and tangent of $\angle\phi$ if ϕ is an acute angle ($0 < \phi < 90°$), and sinϕ = 0.7660.

13. One leg of an isosceles right triangle is 8 cm long.

    a.   How long is the other leg?

    b.   How long is the hypotenuse?

14. In triangle ABC, with right angle at B, the cosine of $\angle A$ is $\frac{7}{18}$. What is the sin $\angle C$?

15. One leg of an isosceles right triangle is 12 in. long. What is the length of the hypotenuse?

16. In triangle PQR, with right angle at P, the sine of $\angle Q$ is 0.235. What is the tangent of $\angle R$?

17.   What is the sign (positive or negative) of sin 290°?

18.   What is the cosine of 240°?

19.   Find the tangent of 330°.

20.   Find the cotangent of a 240° angle in standard position.

21.   Find the secant of a 300° angle in standard position.

22.   If $\theta = \dfrac{4}{5}$, and $\cos\theta$ is negative, what are the values of the five remaining trigonometric functions?

Use the tables that begin on p. 297. Do *not* use a calculator.

23.   Find tan 40.3° to the nearest ten thousandth (four decimal places).

24.   Find sin 71.8° to the nearest ten thousandth (four decimal places).

25.   For what acute angle does sine have a value of 0.5793 if sin 35° = 0.5736 and sin 36° = 0.5878?

Answers and Solutions

1.   ***Answer:*** II   Since the angle is in standard position, its initial side is on the *x*-axis to the right of the origin. the *y*-axis forms a right (90°) angle with the initial side, so a 105° angle's terminal side must sweep past the vertical and into quadrant II.

2.   ***Answer:*** quadrantal   A straight angle in standard position will have its terminal side on the *x*-axis. That makes it a quadrantal angle.

3.   ***Answer:*** no   If angles measuring 245° and 975° were coterminal, then

975° = 245° + *n* · 360°

730° = *n* · 360°

But 730° is *not* a multiple of 360°, so the angles *cannot* be coterminal.

4.   ***Answer:*** 90   A quadrantal angles' terminal side falls on an axis. Therefore, no matter what the size of the coterminal angle, it must be capable of being in the positions of 90°, 180°, 270°, or 0°. All are divisible by 90.

5.   ***Answer:*** None   It is quadrantal. 1260° is three full revolutions from the starting position, and then an additional 180°:

3 · 360° = 1080°

1260° − 1080° = 180°

That places the terminal side on the *x*-axis, left of the origin (negative).

6. **Answer:** 0.5   The sin ratio is $\dfrac{\text{opposite}}{\text{hypotenuse}}$. 5 cm is the length of the side opposite $\angle\alpha$,

   and the hypotenuse is 10 cm. So, $\sin\alpha = \dfrac{5}{10} = \dfrac{1}{2} = 0.5$.

7. **Answer:** $\sqrt{3}$ The tan ratio is $\dfrac{\text{opposite}}{\text{adjacent}}$.

   $5\sqrt{3}$ cm is the length of the side opposite $\angle\beta$, and its adjacent side is 5 cm. So,

   $\tan\beta = \dfrac{5\sqrt{3}}{5} = \sqrt{3}$.

8. **Answer:** 1.25 $\sec\theta = \dfrac{\text{length of hypotenuse}}{\text{length of side adjacent }\theta} = \dfrac{10}{8} = 1.25$

9. **Answer:** 2.5   Cosecant is the reciprocal of sine; that is, $\csc\theta = \dfrac{1}{\sin\theta}$, but you know that

   the value of sin is 0.4; therefore $\csc\theta = \dfrac{1}{0.4}$.

   Divide 1 by 0.4, and you get 2.5.

10. **Answer:** 0.4   Cotangent is the reciprocal function of tangent, so $\cot\theta = \dfrac{1}{\tan\theta}$, but you

    were given tan = 2.5, so $\cot\theta = \dfrac{1}{2.5}$.

    Divide 1 by 2.5 and get 0.4.

11. **Answer:** 0.2632   Sine is cosecant's reciprocal function, $\sin\theta = \dfrac{1}{\csc\theta}$, but you know that

    the value of csc is 3.8; therefore, $\sin\theta = \dfrac{1}{3.8}$.

    Divide 1 by 3.8, and you'll get 0.263157895, which you'll round to 0.2632.

12. **Answer:** 0.6307, 1.2304   First, use the identity $\sin^2\phi + \cos^2\phi = 1$ to find cos ϕ.

    Substitute: $(0.776)^2 + \cos^2\phi = 1$

    Subtract $(0.776)^2$ from both sides: $\cos^2\phi = 1 - (0.776)^2$

    Square the quantity in parentheses: $\cos^2\phi = 1 - 0.6022$

    Subtract 0.6022 from 1: $\cos^2\phi = 0.3978$

    Take the square root of both sides: $\cos^2\phi = 0.6307$

    Now, for the tangent.

    First, write the relevant identity: $\tan\phi = \dfrac{\sin\phi}{\cos\phi}$

    Substitute for sin and cos: $\tan\phi = \dfrac{0.7660}{0.6307}$

    And divide: $\tan\phi = 1.2304$

13. **Answer:** 8 cm, $8\sqrt{2}$ cm

    For part a, both legs of an isosceles right triangle are the same length. As for part b, you have seen that the hypotenuse of an isosceles right triangle is equal in length to a side times the square root of two, but if you forgot that there's always the Pythagorean theorem.

14. **Answer:** $\frac{7}{18}$

    The cofunction identities tell us that in a given right triangle, the sine of one acute angle is the cosine of the other.

15. **Answer:** $12\sqrt{2}$ in.

    In the section "Two Special Triangles," the hypotenuse of an isosceles right triangle is equal to the length of a side times the square root of 2. If you did not recall that, then use the Pythagorean theorem, which—in the case of an isosceles right triangle—may be written $c^2 = a^2 + a^2$ (remember, both legs are equal).

    Substitute: $c^2 = 12^2 + 12^2$

    Square and add: $c^2 = 144 + 144 = 288$

    Solve for c: $c = 12\sqrt{2}$

16. **Answer:** 2.4029   The sine of $\angle Q$ is 0.3842, but you want the tangent of $\angle R$, so you'll use sine's cofunction, $\cos\angle R = 0.3842$. Next, you need to find $\sin \angle R$ so that you may relate sin and cos with the tangent identity.

    First write the equation: $\sin^2 R + \cos^2 R = 1$

    Next, substitute: $\sin^2 R + (0.3842)^2 = 1$

    Clear the parentheses: $\sin^2 R + 0.1476 = 1$

    Collect the constants: $\sin^2 R = 1 - 0.1476$

    Subtract: $\sin^2 R = 0.8524$

    Solve for $\sin R$ $\sin R = 0.9232$

    Now for the tangent identity: $\tan R = \frac{\sin R}{\cos R}$

    Substitute: $\tan R = \frac{0.9232}{0.3842}$

    And divide: $\tan R = 2.4029$

17. **Answer:** negative   290° falls in quadrant IV, where sine is negative. You may want to refer to the figure on p. 40 (Clockwise ACTS, or counter CAST).

18. **Answer:** $-\frac{1}{2}$

    To find the reference angle, subtract 180° from 240°. That's a 60° angle. In a 30-60-90 triangle of hypotenuse 2, the side opposite the 30° angle is 1, and the side opposite the 60° angle is $\sqrt{3}$. That makes $\cos 60° = \frac{1}{2}$, but cosine is negative in the third quadrant, so the answer is $-\frac{1}{2}$.

19. **Answer:** $-\dfrac{\sqrt{3}}{3}$

The angle is in quadrant IV, so subtract it from 360°.

360° − 330° = 30°. Refer back to the 30-60-90 triangle to find the $\tan 30° = \dfrac{1}{\sqrt{3}} = \dfrac{\sqrt{3}}{3}$. In Q-IV, tan is negative; hence the answer is $-\dfrac{\sqrt{3}}{3}$.

20. **Answer:** $\dfrac{\sqrt{3}}{3}$ or 0.577   To find the reference angle, subtract 180°. That makes 60°. Next, if necessary, look back at the figure on p. 36 ("The 30-60-90 right triangle").

$\cot 60° = \dfrac{1}{\sqrt{3}}$, which you can rationalize: $\dfrac{1}{\sqrt{3}} \cdot \dfrac{\sqrt{3}}{\sqrt{3}} = \dfrac{\sqrt{3}}{3}$

Finally, consider that the angle originally terminated in Q-III (third quadrant), where tangent, and therefore cotangent, is positive.

21. **Answer:** 2   Subtract 300° from 360°, and you find that this is a 60° angle in Q-IV.

Secant 60° is $\dfrac{2}{1} = 2$. Since cosine is positive in Q-IV, so is secant.

22. **Answer:** $\cos\theta = -\dfrac{3}{5}$, $\tan\theta = -\dfrac{4}{3}$, $\csc\theta = \dfrac{5}{4}$, $\sec\theta = -\dfrac{5}{3}$, and $\cot\theta = -\dfrac{3}{4}$.

Since $\sin\theta = \dfrac{4}{5}$ and cos θ is negative, the angle must be in the second quadrant. Since in the first quadrant, the cosine would have been positive, and sine is positive in Q-III and Q-IV, the hypotenuse of the triangle must be 5, and the opposite side is 4 (from the sine), which makes the adjacent side 3, either by the Pythagorean theorem, or by remembering the Pythagorean triple, 3-4-5. So, $\cos\theta = -\dfrac{3}{5}$, $\tan\theta = -\dfrac{4}{3}$, $\csc\theta = \dfrac{5}{4}$, $\sec\theta = -\dfrac{5}{3}$, and $\cot\theta = -\dfrac{3}{4}$.

23. **Answer:** 0.8482   40° < 40.3° < 41°

tan 40° = 0.8391, tan 41° = 0.8693.

The difference between them is 1.0724 − 1.0355 = 0.0302.

40.3° is 0.3, or $\dfrac{3}{10}$ of the way between the two angles, so tan 40.3° should be 0.3 of the difference between the two tans: 0.0302 × 0.3 = 0.0091. Since tan 40° is lower than tan 41° add to tan 40°:

0.8391 + 0.0091 = 0.8482

∴ tan 40.3° = 0.8482

24. **Answer:** 0.9499   71° < 71.8° < 72°

sin 71° = 0.9455, sin 72° = 0.9510.

The difference between them is 0.9510 − 0.9455 = 0.0055.

71.8° is 0.8, or $\frac{8}{10}$ of the way between the two angles, so sin 71.8° should be 0.8 of the difference between the two sins: $0.0055 \times 0.8 = 0.0044$. Since sin 71° is lower than sin 72°, add to sin 71°:

$$0.9455 + 0.0044 = 0.9499$$

$$\therefore \sin 71.8° = 0.9499$$

25. **Answer:** 35.4° $1.0 \left\{ x \left\{ \begin{array}{l} \sin 35° = 0.5736 \\ \sin\theta = 0.5793 \\ \sin 36° = 0.5878 \end{array} \right] 0.0057 \right\} 0.0142 \, (x = \text{the difference})$

Now, we'll use the variable x to set up a proportion:

$$\frac{x}{1.0} = \frac{0.0057}{0.0142}$$

$$0.0142x = 0.0057$$

$$x = \frac{0.0057}{0.0142}$$

$$x = 0.4$$

$$\therefore \theta = 35° + 0.4° - 35.4°$$

Supplemental Chapter Problems

Solve these problems for even more practice applying the skills from this chapter. The answer section will direct you to where you need to review.

Problems

1. In which quadrant is the terminal side of a −350° angle?

2. Name four angles that are coterminal with 70°.

3. In which quadrant does the terminal side of an angle of degree measure 1400° fall?

4. What is the lowest possible positive degree measure for an angle that is coterminal with one of −1950°?

5. Name two positive and two negative angles that are coterminal with an angle of degree measure 125°.

Problems 6–7 refer to the following figure.

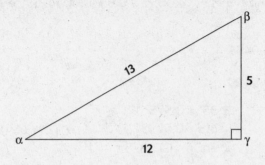

6. Find $\cos\beta$.

7. Find $\csc\alpha$.

8. Find $\cot\beta$.

9. An angle has a cosine of 0.7. What is its secant?

10. An angle has a cosecant of 1.8. Name its inverse function and find its value.

11. Find the sin and tangent of $\angle\lambda$ if λ is an acute angle $(0 < \lambda < 90°)$ and $\cos\lambda = \frac{3}{5}$.

12. Find the sin of $\angle\theta$ if θ is an acute angle $(0 < \theta < 90°)$, $\tan\theta = 2.1445$ and $\cos\theta = 0.4226$.

13. The shortest leg of a 30-60-90 triangle is 10 inches long.

    a.   How long is the other leg?

    b.   How long is the hypotenuse?

14. The hypotenuse of a 30-60-90 triangle is 20 cm long. The shorter leg of the triangle is a, and the longer b. Find the lengths of a and b.

15. In triangle ABC, with right angle at B, the cosine of $\angle A$ is $\frac{12}{13}$. What are the sine of $\angle A$ and the tangent of $\angle A$?

16. What is the sign of tan 225°?

17. What is the sign of csc 310°?

18. What is the cosine of 225°?

19. A 315° angle is in standard position. What is its secant?

20. What is the sine of an 855° angle in standard position?

21. Angle χ is in standard position, and its terminal side passes through the point with coordinates $(-18, 12)$. Find all six of its trigonometric functions.

22. How many seconds are there in an 18° angle?

23. For what acute angle does cosine have a value of 0.7615 if cos 40° = 0.7660 and cos 41° = 0.7547?

24. Find sin 28.7° to the nearest ten thousandth (four decimal places).

25. For what acute angle does sine have a value of 0.8261 if sin 55° = 0.8191 and sin 56° = 0.8290?

Answers

1. I   (Angles and Quadrants, p. 21)

2. 430°, 790°, 1150°, 1510°, and so on.   (Coterminal Angles, p. 25)

3. IV   (Coterminal Angles, p. 25)

4. 150°   (Coterminal Angles, p. 25)

5. −595°, −235°, 485°, 845°, and so on.   (Coterminal Angles, p. 25)

6. $\frac{5}{13}$ (Trigonometric Functions of Acute Angles, p. 28)

7. $\frac{13}{5}$ (Trigonometric Functions of Acute Angles, p. 28)

8. $\frac{5}{12}$ (Reciprocal Trigonometric Functions, p. 29)

9. 1.4286   (Reciprocal Trigonometric Functions, p. 29)

10. sine, 0.5556   (Reciprocal Trigonometric Functions, p. 29)

11. $\frac{4}{5}, \frac{4}{3}$ (Trigonometric Cofunctions, p. 34)

12. 0.9063   (Trigonometric Cofunctions, p. 34)

13. $10\sqrt{3}$ in., 20 in.   (Two Special Triangles, p. 35)

14. 10 cm, $10\sqrt{3}$ cm   (Two Special Triangles, p. 35)

15. $\frac{5}{13}, \frac{5}{12}$ (Functions of General Angles, p. 38)

16. positive   (Functions of General Angles, p. 38)

17. negative   (Functions of General Angles, p. 38)

18. −0.707 or $\frac{\sqrt{2}}{2}$ (Reference Angles, p. 41)

19. 1.4142 or $\sqrt{2}$ (Reference Angles, p. 41)

20. 0.707 or $+\frac{\sqrt{2}}{2}$ (Reference Angles, p. 41)

21.　$\sin = \dfrac{2\sqrt{13}}{13}$, $\cos = \dfrac{-3\sqrt{13}}{\sqrt{3}}$, $\tan = -\dfrac{2}{3}$, $\csc = \dfrac{\sqrt{13}}{2}$, $\sec = -\dfrac{\sqrt{13}}{3}$, and $\cot = -\dfrac{3}{2}$

　　　(Reference Angles, p. 41)

22.　64,800　　(Trig Tables versus Calculators, p. 44)

23.　40.4°　　(Interpolation, p. 45)

24.　0.4802　　(Interpolation, p. 45)

25.　55.7°　　(Interpolation, p. 45)

Chapter 2
Graphs of Trigonometric Functions

This chapter actually deals with a lot more than the graphs of the trigonometric functions, but we had to name it something. Eventually, we'll get to what it says in the title, but there's a lot of ground to cover on the way there.

Understanding Degree Measure

We have already discussed measuring angles by degree of rotation, but we need to point out here that that's only one way to measure angles. For rough estimates of the degree measure of angles, you can use an instrument called a *protractor*. No doubt you have seen a protractor of one type or another at some time in your school career, but one type is pictured here.

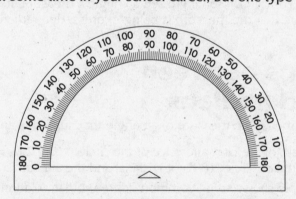

A protractor used to find degree measure of angles.

A protractor is also useful for drawing angles, as long as you're dealing with whole degrees. For fractional angles, or minutes and seconds, a protractor is useless. But, degrees are not the only way that angles may be measured. Not as common in the nonmathematical world, but quite popular with trigonometry teachers and students, are **radians.** In trigonometry, not only are radians equally as important as degrees, but it is essential that you know how to convert from one unit to the other.

Understanding Radians

If a central angle of a circle **subtends** (cuts off) an arc whose length is $\frac{1}{360}$ of the circle's circumference, then that angle has an angle measure of 1°. If a central angle subtends an arc that is equal in length to the circle's radius, then the measure of that angle is 1 radian. Look at the following figure to see what an angle of 1 radian looks like.

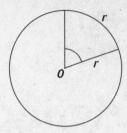

An angle of measure 1 radian.

If angle ϕ subtends an arc of s centimeters, then we can define its angle measure as $\phi = \dfrac{s \text{ cm}}{r \text{ cm}}$.

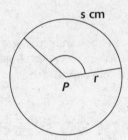

A central angle with arc length s cm.

Since both numbers are in the same unit, the units cancel $\phi = \dfrac{s \text{ cm}}{r \text{ cm}} = \dfrac{s}{r}$. So radian measure is expressed with no units. For that reason if we see an angle's measure stated with no units written after it, we can assume that the measure is being expressed in radians.

Relationships between Degrees and Radians

What do you suppose is the radian measure of an angle that subtends an arc of 360°?

An arc of 360° is the length of the circumference of the circle, πd or $2\pi r$. How convenient that is! That means that an arc of half a circle, $180° = \pi$ radians, and the arc subtended by a right angle, $90° = \dfrac{\pi}{2}$. The following table shows some common degree/radian relationships:

Some Common Degree/Radian Values									
degrees	0°	30°	45°	60°	90°	120°	130°	150°	180°
radians	0	$\dfrac{\pi}{6}$	$\dfrac{\pi}{4}$	$\dfrac{\pi}{3}$	$\dfrac{\pi}{2}$	$\dfrac{2\pi}{3}$	$\dfrac{3\pi}{4}$	$\dfrac{5\pi}{6}$	π

Example Problems

These problems show the answers and solutions.

1. What is the radian measure of a 50° angle?

    answer: **0.873**

Set up a proportion based upon the fact that $180° = \pi$ radians:

$$\frac{50}{180} = \frac{x}{\pi}$$

$$180x = 50\pi$$

$$x = \frac{50\pi}{180} = \frac{5\pi}{18} = 0.873$$

2. What is the degree measure of an angle of $\frac{\pi}{8}$?

    answer: **22.5°**

    You are looking to express $\frac{\pi}{8}$ radians, but, remember that $180° = \pi$ radians, so

    $$\frac{\pi}{8} = \frac{180°}{8} = 22.5°$$

    You also could have solved the proportion: $\quad \frac{\frac{\pi}{8}}{\pi} = \frac{x}{180}$

    $$\pi x = \frac{180\pi}{8} = 22.5\pi$$
    $$x = 22.5°$$

3. What is the radian measure of a 270° angle?

    answer: $\frac{3\pi}{2}$

    First the long way: $\quad \frac{270°}{180°} = \frac{\theta}{\pi}$

    $$\frac{\cancel{270}^{\,3}}{\cancel{180}_{\,2}} = \frac{\theta}{\pi}$$

    $$2\theta = 3\pi$$

    $$\theta = \frac{3\pi}{2}$$

    Now the short way: You know that $270° = 180° + 90°$, and you know from the previous table that

    $$180° = \pi \text{ and } 90° = \frac{\pi}{2}.$$

    Add them:

    $$\pi + \frac{\pi}{2} = \frac{2\pi}{2} + \frac{\pi}{2} = \frac{3\pi}{2}$$

4. What is the length of an arc subtended by an angle of 1.5 radians in a circle with a diameter of 10 inches?

    answer: **7.5 inches**

    An arc length of one radian is equal to the radius of the circle. This circle's radius is half the diameter, or 5 inches. $1.5 \times 5 = 7.5$ inches.

The Unit Circle and Circular Functions

The graph of the equation $x^2 + y^2 = 1$, when graphed on the Cartesian (or rectangular) coordinate system is a circle. This graph has its center at the origin and has a radius of 1 unit.

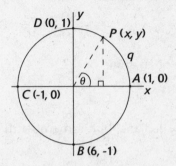

The unit circle ($x^2 + y^2 = 1$).

A function has a dependent variable y corresponding to each value for the independent variable, x. For a relationship to be a true function, each value of x must have one and only one y associated with it. A quick glance at the above figure reveals whether or not a circle is a function. Look at points B and D. Both have an **abcissa** (x-value) of 0, but D's **ordinate** (y-value) is 1, while B's ordinate is -1. Is a circle a function? Obviously not.

Trigonometric functions, like circles, are relationships rather than functions, but traditionally, mathematicians have winked and referred to these relationships as functions. Circular functions are a new and different way of confirming the trigonometric functions. They don't need to refer to any angle, although the central angle, θ, is obviously there. Circular functions use radians and refer to the arc, in the case of the preceding figure, arc q.

Since the radius of the circle is 1, we can readily find the value of $\sin q$ in the above figure:

$$\sin q = \frac{y}{1} = y$$

If you're not sure how we got that, it's okay to refer back to the trigonometric reference to $\frac{\text{opposite}}{\text{hypotenuse}}$, or to remember SOHCAHTOA.

Now let's find the value of $\cos q$ in the above figure: $\cos q = \frac{x}{1} = x$.

Sine and cosine are considered the two basic circular functions, and they serve to find the values of all the other functions. Tangent is $\frac{\sin}{\cos}$, so $\tan q = \frac{\sin q}{\cos q} = \frac{y}{x}$; $x \neq 0$.

As for the inverse functions,

$$\csc q = \frac{1}{\sin q} = \frac{1}{y};\ y \neq 0$$

$$\sec q = \frac{1}{\cos q} = \frac{1}{x};\ x \neq 0$$

$$\cot q = \frac{\cos q}{\sin q} = \frac{x}{y};\ y \neq 0$$

Notice that with all functions where the denominator might be a zero, it is specified that the denominator may not be zero, since if it were, the function would be undefined.

Domain versus Range

Referring to the preceding figure, the coordinates of point P ($\cos q$, $\sin q$) correspond to an arc with a length of $|q|$. Since this length can be positive (counterclockwise) or negative (clockwise) the *domain* (x-values) of each of these functions is the set of real numbers. The *range* (y-values), however, is considerably more limited. The sine and cosine are the abcissa and ordinate of a point moving around the circumference of a unit circle, and they can be anywhere between -1 and 1, so the range of these two functions are such that $-1 \leq r \leq 1$ (be sure to look at the preceding figure).

Example Problems

These problems show the answers and solutions.

1. What values of x in the domain of the sine function between $-\pi$ and π have a range equal to an absolute value of 1?

    answer: $-\dfrac{\pi}{2}, \dfrac{3\pi}{2}$

    Rather than trying to think in radian terms, think in the unit you're more accustomed to—degrees. The sine function reaches $|1|$ twice between $-\pi$ and π, positive 1 at 90°, and -1 at 270°.

    Those points have x-coordinates $-\dfrac{\pi}{2}$ and $\dfrac{3\pi}{2}$. Check out the following figure.

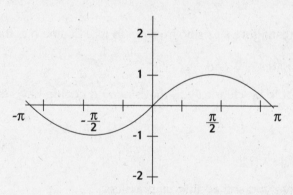

Sine from $-\pi$ to π.

2. What values of x in the domain of the cosine function between $-\pi$ and π have a range equal to an absolute value of 1?

    answer: $-\pi$, **0**, π

    Again, rather than trying to think in radian terms, think in the more familiar degrees. The cosine function reaches $|1|$ three times between $-\pi$ and π: positive 1 at 0° and -1 at $-180°$ and 180°. Those points have x-coordinates $-\pi$, 0, and π. Check out the following figure.

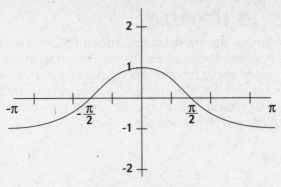

Cosine from −π to π.

3. What values of x in the domain of the sine function between −π to π have a range equal to 0?

    answer: −π, **0**, π

    Look back at the figure previous to this one ("Sine from −π to π"). Sine hits the x-axis at 0°, 180°, and 270°. Those are x-coordinates −π, 0, and π.

4. What values of x in the domain of the cosine function between −π and π have a range equal to 0?

    answer: $-\dfrac{\pi}{2}, \dfrac{\pi}{2}$

    Look back at the latest figure ("Cosine from −π to π"). Cosine hits the x-axis at 90° and 270°.

    Those are x-coordinates $-\dfrac{\pi}{2}$ and $\dfrac{\pi}{2}$.

From the problems, it should be clear to you that sine and cosine have exactly the same graphs, only 90° out of phase.

Work Problems

Use these problems to give yourself additional practice.

1. What is the radian measure of a 75° angle?

2. What is the degree measure of an angle of $\dfrac{\pi}{4}$?

3. What is the radian equivalent of a 225° angle?

4. What is the length of an arc subtended by an angle of 1.25 radians in a circle with a diameter of 24 cm?

5. What values in the domain of the sine function between −2π and 2π radians have a range of 1?

Worked Solutions

1. **1.309** Set up a proportion based upon the fact that $180° = \pi$ radians:

$$\frac{75}{180} = \frac{x}{\pi}$$

$$180x = 75\pi$$

$$x = \frac{75\pi}{180} = \frac{5\pi}{12} = 1.309$$

2. **45°** We are looking to express $\frac{\pi}{4}$ radians, but, remember that $180° = \pi$ radians, so:

$$\frac{\pi}{4} = \frac{180°}{4} = 45°$$

3. $\frac{5\pi}{4}$ First the long way:

Make a proportion: $\frac{225°}{180°} = \frac{\theta}{\pi}$

Divide 15 out of both terms $\frac{\cancel{225}}{\cancel{180}} \rightarrow \frac{5}{4} = \frac{\theta}{\pi}$

$4\theta = 5\pi$

$\theta = \frac{5\pi}{4}$

Or you could have used the short way. You know that $225° = 180° + 45°$, and you know (looking back to the table on page 60) that $180° = \pi$ and $45° = \frac{\pi}{4}$ radians.

Add them:

$\pi + \frac{\pi}{4} = \frac{4\pi}{4} + \frac{\pi}{4} = \frac{5\pi}{4}$

4. **15 cm** An arc length of one radian is equal to the radius of the circle. This circle's radius is half the diameter, or 12 cm.

$1.25 \times 12 = 15$ cm

5. $-\frac{3\pi}{2}, \frac{\pi}{2}$ It would be most helpful if you had a graph of the sine curve from -2π to 2π, like in the following figure.

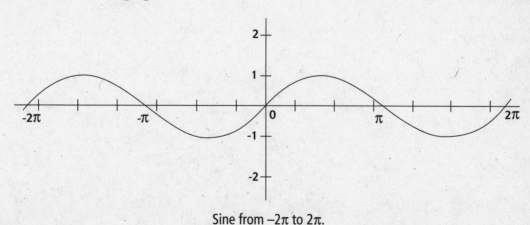

Sine from -2π to 2π.

As you can see, the leftmost peak of 1 is at x-coordinate halfway between $-\pi$ and -2π or $-\frac{3\pi}{2}$.

The second one occurs halfway between 0 and π, or at $\frac{\pi}{2}$.

Periodic Functions

Remember the unit circle? It is drawn again in the following figure.

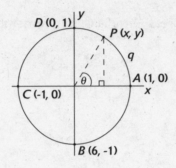

A unit circle.

Since the circumference of a circle is $2\pi r$, and r in the unit circle is 1, the circumference of a unit circle is 2π. Point P is the terminating point of central angle θ, which is equal to q radians. Furthermore, if a length of 2π is added to q radians, its terminating point would also be P; so would an angle 4π bigger, 6π bigger, 8π bigger, and so forth. In other words, point P is the terminating point for an infinite number of angles, both positive and negative, each of which is a multiple of two times π greater or smaller than angle θ. You may think of point P as being a bicycler riding around a circular track. Beginning at P, every time he completes a 2π distance around the track, he will return to point P. The same is true of every other point on the circle as well. Each is a terminal point for an infinite number of angles.

Functions that have this property are called **periodic functions.** A function f is periodic if there is a positive real number, k, such that $f(x + k) = f(x)$ for all x values in f's domain. The smallest value of k for which this is true is known as f's **period.** A function's period may also be described as the distance between repetitions. The periods of the sine and cosine functions are each 360°, or 2π radians.

The following are true of periodic functions:

$\sin(\theta + 2\pi) = \sin\theta$
$\cos(\theta + 2\pi) = \cos\theta$
$\sin(\theta + 360°) = \sin\theta$
$\cos(\theta + 360°) = \cos\theta$

If *n* is an integer, then:

$$\sin(\theta + 2n\pi) = \sin\theta$$
$$\cos(\theta + 2n\pi) = \cos\theta$$
$$\sin(\theta + 360n°) = \sin\theta$$
$$\cos(\theta + 360n°) = \cos\theta$$

All of the preceding would also be true if the + signs were changed to − signs. From studying the periodic properties of circular functions, the solutions to many real-world problems may be found. Planetary motion, sound waves, earthquake waves, tidal motion, and electric current generation all involve periodic motion

Example Problems

These problems show the answers and solutions.

1. If $\sin y = \frac{\sqrt{2}}{5}$, what is $\sin(y + 8\pi)$, and $\sin(y + 4\pi)$?

    answer: $\frac{\sqrt{2}}{5}$, $\frac{\sqrt{2}}{5}$

    Sin is a periodic function, whose period is 2π. Adding any multiple of 2π to it will not change its value, since it returns you to the same point on the circle.

2. If $\sin y = \frac{\sqrt{3}}{7}$, what is $\sin(y − 6\pi)$, and $\sin(y − 2\pi)$?

    answer: $\frac{\sqrt{3}}{7}$, $\frac{\sqrt{3}}{7}$

    Sin is a periodic function, whose period is 2π. Subtracting any multiple of 2π from it will not change its value, since it returns you to the same point on the circle.

3. If $\cos y = \frac{3}{8}$, what is $\cos(y + 6\pi)$, and $\cos(y − 10\pi)$?

    answer: $\frac{3}{8}$, $\frac{3}{8}$

    Cosine is a periodic function, whose period is 2π. Adding or subtracting any multiple of 2π will not change its value, since it returns you to the same point on the circle.

Graphing Sine and Cosine

We've already looked at the graphs of sine and cosine, but let's look at a single period of each from 0 to 2π radians (360°). First let's find some sine and cosine values for different points in one period. Flip the page and rotate your book clockwise.

Values of Sine and Cosine Functions at Several Different Points

radians	0	$\frac{\pi}{6}$	$\frac{\pi}{4}$	$\frac{\pi}{3}$	$\frac{\pi}{2}$	$\frac{2\pi}{3}$	$\frac{3\pi}{4}$	$\frac{5\pi}{6}$	π	$\frac{7\pi}{6}$	$\frac{5\pi}{4}$	$\frac{4\pi}{3}$	$\frac{3\pi}{2}$	$\frac{5\pi}{3}$	$\frac{7\pi}{4}$	$\frac{11\pi}{6}$	2π
degrees	0°	30°	45°	60°	90°	120°	135°	150°	180°	210°	225°	240°	270°	300°	315°	330°	360°
$\sin\theta$	0	0.5	0.707	0.866	1	0.866	0.707	0.5	0	-0.5	-0.707	-0.866	-1	-0.866	-0.707	-0.5	0
$\cos\theta$	1	0.866	0.707	0.5	0	-0.5	-0.707	-0.866	-1	-0.866	-0.707	-0.5	0	0.5	0.707	0.866	1

Now, plot those values on a graph to get the following graph for one period of sine and cosine:

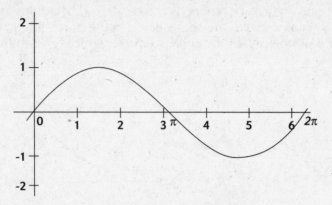

One period of the sine function.

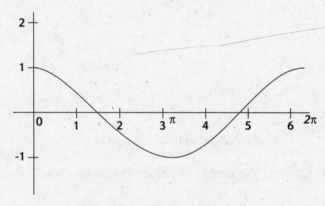

One period of the cosine function.

Since both sine and cosine are periodic functions with a period of 2π, the graph would repeat it-self to the left and the right of the shown portion infinitely. We noted earlier that cosine and sine are 90° ($\frac{\pi}{2}$ radians) out of phase with each other. This is easily shown by drawing both on the same pair of axes, as has been done in the following figure.

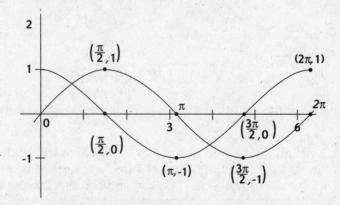

One period of both sine and cosine functions.

Notice the marked points on this figure. When sine is at (0,0), cosine is at (0,1). 90° later, at ($\frac{\pi}{2}$, 0), sine is up to 1, and cosine is down to 0. Check the full pattern over the course of the whole period, and you'll see that if cosine's graph were slid $\frac{\pi}{2}$ to the right, it would coincide with the sine's graph.

Vertical Displacement and Amplitude

The appearance of the sine and cosine functions can be altered by adding terms to the basic equation, $y = \sin x$ or $y = \cos x$, which are represented in the preceding graphs. Adding a constant, a, to the equation of sine causes the curve to move vertically, as in the following figure.

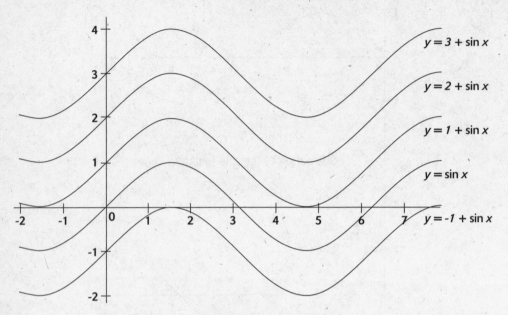

Some vertical shifts of the sine function.

The same applies to the graph of the cosine function, as you can see in the figure that follows.

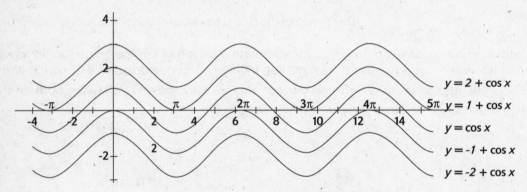

Some vertical shifts of the cosine function.

Notice what the magnitude of the shift is that's brought about by each number added to the sine or cosine.

The distance a periodic function travels from the x-axis is known as its **amplitude.** The functions $y = \sin x$ and $y = \cos x$ travel to a maximum height of 1 and a minimum height of -1. That means that the amplitude of $y = \sin x$ or $y = \cos x$ is 1. You can arrive at this conclusion by adding together the absolute values of the highest and lowest points and dividing by 2. Multiplying sine or cosine by a constant, b, will change the amplitude of the graphs.

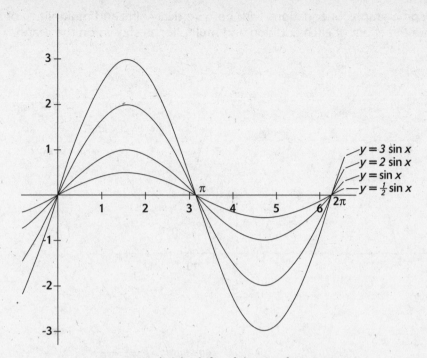

Some amplitude shifts of the sine function.

The same applies to the graph of the cosine function, as you can see in the following figure.

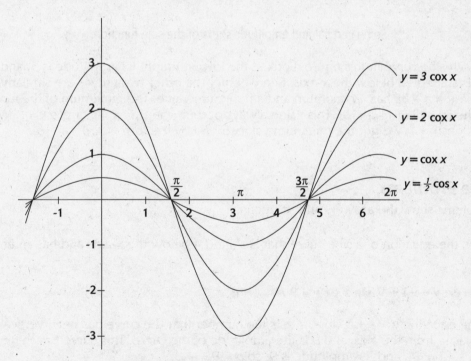

Some amplitude shifts of the cosine function.

Notice the magnitude of the difference in amplitude that results from a multiplier of each specified size.

Now let's see some graphs of equations with both vertical shifts and multipliers, of the form $y = a + bx$. Note the effect of each addition and multiplier as shown on the graph.

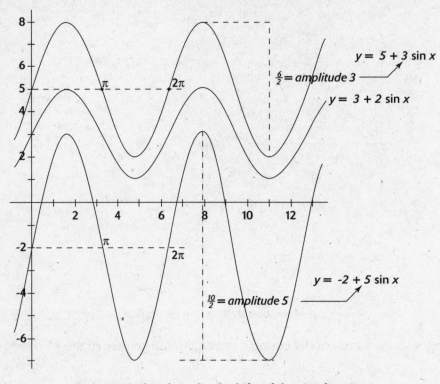

Some vertical and amplitude shifts of the sine function.

Examine each sine curve in this figure. Look at the lowest graph. Its amplitude is 5, and its vertical center is shifted −2 below the x-axis. See whether the other two curves are similarly affected. The function $y = a + bx$ has a maximum and a minimum value. The maximum of the function is found by the formula $M = a + |b|$. This value always occurs when sin $x = 1$ or cos $x = 1$. When those functions $= -1$, we get the minimum value, found by the formula $m = a - |b|$.

Example Problems

These problems show the answers and solutions.

1. Write the equation of a sine curve that is shifted 4 below the x-axis and has an amplitude of 9.

    answer: $y = -4 + 9 \sin x$ **or** $y = 9 \sin x - 4$

    In the equation $y = a + b \sin x$, a tells the number that the curve has been vertically displaced from the axis, and b is the amplitude of the curve. This curve has shifted down four, so $a = -4$, and its amplitude is 9, so $b = 9$.

2. Write the equation of a cosine curve that is shifted 5 above the x-axis and has an amplitude of 4. What are its maximum and minimum values?

   ***answer:** y = 5 + 4 cos x, M = 9, m = 1*

   In the equation $y = a + b \sin x$, a tells the number that the curve has been vertically displaced from the axis, and b is the amplitude of the curve. This curve has shifted up five, so $a = 5$, and its amplitude is 4; so $b = 4$; $M = 5 + 4 = 9$; $m = 5 - 4 = 1$.

3. Write the equation of a sine curve that is shifted $\frac{1}{2}$ space below the x-axis and has a high point of $3\frac{1}{2}$ and a low point of $-4\frac{1}{2}$.

   ***answer:** y = -\frac{1}{2} + 4 \sin x$ or $y = 4 \sin x - \frac{1}{2}*

   In the equation $y = a + b \sin x$, a tells the number that the curve has been vertically displaced from the axis, and b is the amplitude of the curve. This curve has shifted down one half, so $a = -\frac{1}{2}$, and its amplitude is half the sum of $\left|3\frac{1}{2}\right| + \left|-4\frac{1}{2}\right| = 8$, so $b = 4$.

Work Problems

Use these problems to give yourself additional practice.

1. If $\sin w = \frac{\sqrt{3}}{5}$, what is $\sin(w + 12\pi)$?

2. If $\cos y = \frac{7}{16}$, what is $\cos(y - 8\pi)$?

3. Write the equation of a cosine curve that is shifted 3 units down and has an amplitude of 6. What are its maximum and minimum values?

4. Write the equation of a sine curve that is shifted 2 units down and has a high point of 5 and a low point of −7.

5. Write the equation of a cosine curve that is shifted $\frac{1}{2}$ units up and has a high point of $5\frac{1}{2}$ and a low point of $-4\frac{1}{2}$.

Worked Solutions

1. $\frac{\sqrt{3}}{5}$ Sin is a periodic function, whose period is 2π. Adding any multiple of 2π to it will not change its value, since it returns you to the same point on the circle.

2. $\frac{7}{16}$ Cosine is a periodic function, whose period is 2π. Adding or subtracting any multiple of 2π will not change its value, since it returns you to the same point on the circle.

3. **$y = -3 + 6 \cos x$ or $y = 6 \cos x - 3$, $M = 3$; $m = -9$**

   In the equation $y = a + b \sin x$, a tells the number that the curve has been vertically displaced from the axis, and b is the amplitude of the curve. This curve has shifted down three, so $a = -3$, and its amplitude is 6, so $b = 6$. That makes $M = (-3) + 6 = 3$ and $m = -3 - 6 = -9$.

4.   **$y = -2 + 6\sin x$ or $y = 6\sin x - 2$**

In the equation $y = a + b\sin x$, a tells the number that the curve has been vertically displaced from the axis, and b is the amplitude of the curve. This curve has shifted down 2, so $a = -2$, and its amplitude is half the sum of $|5| + |-7| = 12$, so $b = 6$.

5.   **$y = \dfrac{1}{2} + 5\cos x$**   In the equation $y = a + b\cos x$, a tells the number that the curve has been vertically displaced from the axis, and b is the amplitude of the curve.

This curve has shifted up one half, so $a = a = \dfrac{1}{2}$, and its amplitude is half the sum of $\left|5\dfrac{1}{2}\right| + \left|-4\dfrac{1}{2}\right| = 10$, so $b = 5$.

Frequency and Phase Shift

Two other variations are possible in graphing the sine and cosine functions. The first deals with the addition of c in the position shown here: $y = \sin cx$. This multiplier affects the **frequency** of the curve—the number of times the curve appears in $360°$, or 2π radians.

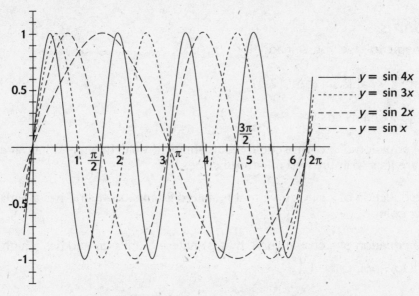

Some different frequencies of the sine function.

Notice in this figure that the number of times the entire sine curve repeats is determined by the coefficient of x. That is, where $c = 2$, there are 2 whole sine curves, where $c = 3$, there are 3 whole sine curves, and so on. The same is true of the cosine curve, as you can see in the following figure.

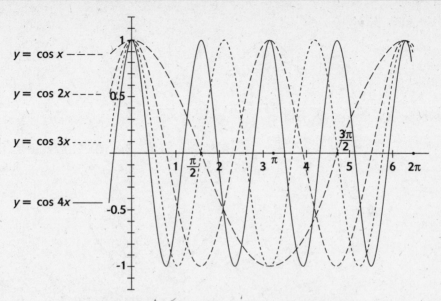

Some different frequencies of the cosine function.

We've commented before on the cosine's graph's being the same as the sine's but with a 90° **phase shift.** Now we'll show you how a phase shift is accomplished. Look at the equation $y = \sin (x + d)$. The added d inside the parentheses serves to move the graph to the left if positive, or to the right if negative.

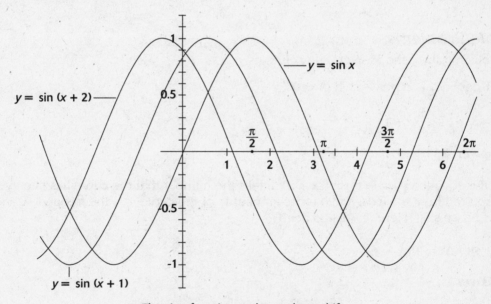

The sine function and two phase shifts.

Note in the above figure that the graph of $y = \sin x$ crosses the x-axis at 0; $y = \sin (x + 1)$ crosses at −1; and $y = \sin (x + 2)$ crosses it at −2. The following figure shows two phase shifts the other way for cosine.

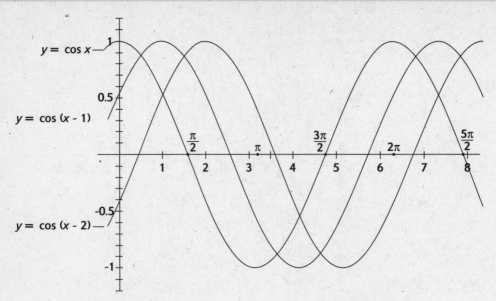

The cosine function and two phase shifts in the other direction.

To turn a sine curve into a cosine or vice versa, just do this:

$$\sin x = \cos\left(x - \frac{\pi}{2}\right) \text{ or } \cos x = \sin\left(x + \frac{\pi}{2}\right)$$

Example Problems

These problems show the answers and solutions.

Find the frequency and phase shift of each.

1. $y = 3 + \frac{1}{2}\sin(2x + 1)$

   answer: 2, 1

   In the equation $y = a + b\sin(cx + d)$, a tells the number that the curve has been vertically displaced from the axis, and b is the amplitude of the curve; c is the frequency, and d is the phase shift. Here, $c = 2$ and $d = 1$.

2. $y = \sin(3x - 4)$

   answer: 3, −4

   In the equation $y = a + b\sin(cx + d)$, a tells the number that the curve has been vertically displaced from the axis, and b is the amplitude of the curve; c is the frequency, and d is the phase shift. Here, $c = 3$ and $d = -4$.

3. $y = 4 - 3\cos(2x - 7)$

answer: 2, –7

In the equation $y = a + b\cos(cx + d)$, a tells the number that the curve has been vertically displaced from the axis, and b is the amplitude of the curve; c is the frequency, and d is the phase shift. Here, $c = 2$ and $d = -7$.

Graphing Tangents

Remember that $\tan\theta = \dfrac{\sin\theta}{\cos\theta}$. Whenever cosine = 0, tangent is undefined. This occurs every $\dfrac{n\pi}{2}$, where n is an odd integer. At these points, tangent's value approaches infinity.

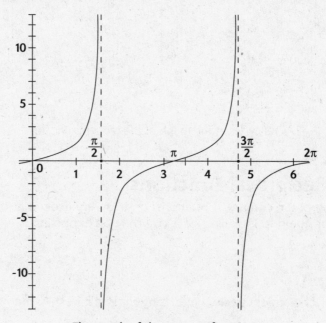

The graph of the tangent function.

Asymptotes

You can see the tangent curve approach infinity at $\dfrac{\pi}{2}$ and at $\dfrac{3\pi}{2}$.

Notice the dotted lines. It is customary to draw these lines when graphing the tangent function. They are called **asymptotes** (pronounced ASS-im TOTES) and are values that the tangent graph approaches very closely as it gets very large, but never actually reaches. You should also notice that one period of the tangent function is π radians, or 180°.

From the graph, you should observe that as the tangent moves in a positive direction, the graph is always climbing. Beginning at zero, it climbs to ∞ and approaches its asymptote until it has no value—the asymptote is never reached. Disappearing, it comes back on the other side of the asymptote at –∞, climbs through zero to ∞, disappears again, and then reappears at –∞, climbing to zero. That was two full periods in 360°. The figure that follows shows several cycles of the tangent function.

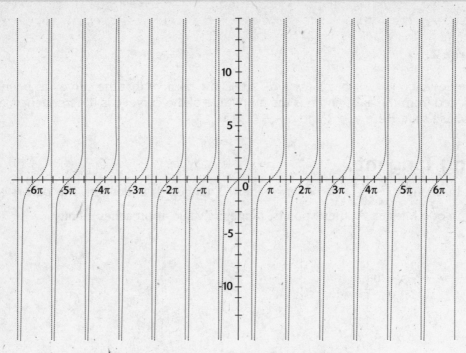

Several cycles of the graph of the tangent function.

Graphing the Reciprocal Functions

With the possible graph of the cotangent relationship, it is very likely that the graphs of the reciprocal functions are not going to look like what you may be expecting.

Cotangent's Graph

Let's look at the graph of cotangent while that of tangent is still fresh in our minds (and just above), for easy reference.

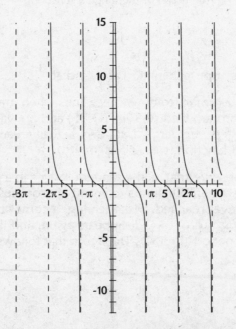

Several cycles of the graph of the cotangent function.

It has almost every characteristic of tangent, including asymptotes, except that it, everywhere, is running from higher to lower. Since both tangent and cotangent are boundless in the vertical direction, they have no defined amplitudes. The general forms of the tangent and cotangent functions are $y = a + b \tan (cx + d)$ and $y = a + b \cot (cx + d)$.

The variables c and d, as in the case of the sine and cosine functions, determine the period and phase shift of the functions. The period is $\frac{\pi}{c}$, and the phase shift is $\frac{d}{c}$. If the value of $\frac{d}{c}$ is greater than 0, the shift is to the left. A negative value for $\frac{d}{c}$ shifts the graph to the right. The variable a represents a vertical shift, and since there is no amplitude, b represents how much the graph may be stretched out in a vertical direction.

We can find the asymptotes by solving $cx + d = \frac{\pi}{2}$ and $cx + d = -\frac{\pi}{2}$ for x.

Example Problems

These problems show the answers and solutions.

All of the problems in this section refer to the function, $y = \tan\left(\frac{\pi}{6}x + \frac{\pi}{12}\right)$.

1. Find the function's graph's asymptotes.

    answer: 5, $-\frac{7}{2}$ or $-3\frac{1}{2}$

    To find the asymptotes, solve these for x:

    $$cx + d = \frac{\pi}{2} \quad \text{and} \quad cx + d = -\frac{\pi}{2}$$

    $$\frac{\pi}{6}x + \frac{\pi}{12} = \frac{\pi}{2} \quad \text{and} \quad \frac{\pi}{6}x + \frac{\pi}{12} = -\frac{\pi}{2}$$

    $12 \times$ everything clears denominators:

    $$12\left(\frac{\pi}{6}x + \frac{\pi}{12} = \frac{\pi}{2}\right) \qquad 12\left(\frac{\pi}{6}x + \frac{\pi}{12} = -\frac{\pi}{2}\right)$$

    Multiply: $2\pi x + \pi = 6\pi \qquad 2\pi x + \pi = -6\pi$

    Collect terms: $\pi x = 6\pi - \pi \qquad \pi x = -6\pi - \pi$

    Now add: $\pi x = 5\pi \qquad\qquad 2\pi x = -7\pi$

    . . . and divide: $x = 5 \qquad\qquad x = -\frac{7}{2}$

2. Find the function's graph's phase shift.

    answer: $\frac{1}{2}$

    Phase shift is $\frac{d}{c}$, which may be rewritten as $d \div c$.

    To solve: $d \div c = \frac{\pi}{12} \div \frac{\pi}{6}$

    That becomes reciprocal multiplication: $\frac{\pi}{12} \div \frac{\pi}{6} = \frac{\pi}{12} \times \frac{6}{\pi} = \frac{6\pi}{12\pi} = \frac{1}{2}$

3. Find the function's graph's amplitude.

   answer: **not defined**

   The tangent graph has infinite height, and so has no defined amplitude.

Graphs of Secant and Cosecant

Cosecant and sine are drawn on the same set of axes, from -2π to 2π.

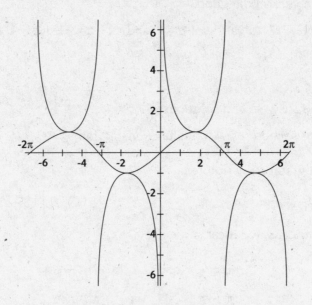

Sine and cosecant functions drawn on the same axes.

Notice that although sine is a continuous function, cosecant is not. There are asymptotes with cosecant's domain, and the curve's range goes everywhere sine does not, except at 1 and −1.

Now look at the graphs of secant and cosine.

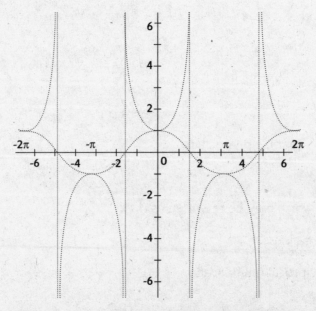

Cosine and secant functions drawn on the same axes.

We should not be surprised that secant's and cosecant's graphs are identical, but out of phase by $\frac{\pi}{2}$. Everything said about cosecant applies to secant. Both have the same period as sine and cosine, 2π. The phase shift of each is found by solving the equation $cx - d = 0$ for x, while the period is found by solving $cx = 2\pi$ for x. For asymptotes, solve $cx + d = 0$, $cx + d = \pi$, and $cx + d = 2\pi$ for x.

The following four principles are true of reciprocal functions, and, for that matter of reciprocals in general.

- ❏ A real number and its reciprocal move in opposite directions. That is to say, as $\sin x$ increases, $\csc x$ is decreasing. This is also true of $\cos x$ and $\sec x$, and of $\tan x$ and $\cot x$.

- ❏ As a function approaches a value of 0, its reciprocal approaches positive or negative infinity— that is, it increases or decreases without limit. The converse is also true.

- ❏ The reciprocal of 1 is 1, and the reciprocal of −1 is −1. You can see this in the case of the functions $\sin 90°$ and $\csc 90°$, both of which are equal to 1, or the cos and secant of 0°, both of which equal 1. Similarly, sin and csc 270° and cos and sec 180° all equal −1.

- ❏ A real number and its reciprocal always have the same sign, so whatever quadrant the tangent is positive in, the cotangent will also be positive in, and so forth.

Example Problems

These problems show the answers and solutions.

All of the problems in this section refer to the function, $y = 2\csc\left(\frac{\pi}{2}x + \frac{\pi}{2}\right)$.

1. Determine where the function's asymptotes are located.

   answer: **−1, −3, −5**

$cx + d = 0$	$cx + d = \pi$	$cx + d = 2\pi$
$\frac{\pi}{2}x + \frac{\pi}{2} = 0$	$\frac{\pi}{2}x + \frac{\pi}{2} = \pi$	$\frac{\pi}{2}x + \frac{\pi}{2} = 2\pi$
$\frac{\pi}{2}x = -\frac{\pi}{2}$	$\frac{\pi}{2}x = -\frac{3\pi}{2}$	$\frac{\pi}{2}x = -\frac{5\pi}{2}$
$x = -1$	$x = -3$	$x = -5$

2. Find the function's phase shift.

   answer: **1**

   $cx - d = 0$

   $\frac{\pi}{2}x - \frac{\pi}{2} = 0$

   $\frac{\pi}{2}x = \frac{\pi}{2}$

   $x = \dfrac{\frac{\pi}{2}}{\frac{\pi}{2}} \rightarrow\rightarrow x = \frac{\pi}{2} \times \frac{2}{\pi}$

   $x = 1$

3. Find the function's period.

   answer: 4

   $cx = 2\pi$

   $\dfrac{\pi}{2}x = 2\pi$

   $x = \dfrac{2\pi}{\frac{\pi}{2}} \rightarrow \rightarrow x = \dfrac{2\pi}{2} \times \dfrac{2}{\pi}$

   $x = 4$

Work Problems

Use these problems to give yourself additional practice.

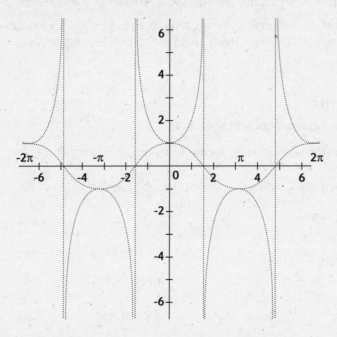

1. Which trigonometric function's graphs are shown in this figure?

2. As a function approaches infinity, where is its reciprocal?

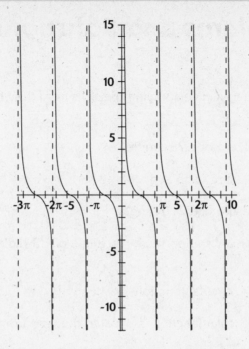

3.   Which trigonometric function's graph is shown in this figure?

4.   Find the phase shift of the function $y = a + b \csc(cx - d)$.

Worked Solutions

1.   **cosine, secant**   Notice that the periodic function varies between 1 and −1 and is sinusoidal (that means "sine like"). That makes it either sine or cosine. Since its y-value at $x = 0$ is 1, it must be cosine. The discontinuous curve is either secant or cosecant. Since it coincides with the cosine curve only at its turning points, it must be cosine's reciprocal curve, secant.

2.   **approaching zero**   As a function approaches zero, its reciprocal is approaching positive or negative infinity. The converse, then, must also be true.

3.   **cotangent**   It's discontinuity and asymptotes, as well as its shape should tell you that this is either tangent or cotangent, but how do we know which one? Moving from left to right on the graph, tangent is always rising. Cotangent falls from left to right. This graph falls from left to right, so it must be cotangent.

4. $\dfrac{-d}{c}$ Phase shift is always found by solving $cx - d = 0$ for x, or simply by dividing the d term by the c term. Since the d term is $-d$, the phase shift is $\dfrac{-d}{c}$.

Chapter Problems and Solutions

Problems

Solve these problems for more practice applying the skills from this chapter. Worked out solutions follow problems.

1. What is the radian measure of a 75° angle?

2. What is the degree measure of an angle of $\frac{\pi}{6}$?

3. What is the radian measure of a 210° angle?

4. What is the length of an arc subtended by an angle of 1.3 radians in a circle with a diameter of 12 cm?

5. Which are the two basic circular functions?

6. In a unit circle, which circular function is equal to the y-coordinate?

7. What values of x in the domain of the sine function between -2π and 2π have a range equal to an absolute value of 1?

8. What values of x in the domain of the sine function between -2π and 2π have a range equal to 0?

9. If $\sin y = \frac{\sqrt{5}}{8}$, what is $\sin(y - 10\pi)$ and $\sin(y - 12\pi)$?

10. If $\cos y = \frac{1}{2}$, what is $\cos(y + 4\pi)$ and $\cos(y - 20\pi)$?

11. Write the equation of a sine curve that is shifted 6 units down and has an amplitude of 11.

12. Write the equation of a cosine curve that is shifted 7 units up and has an amplitude of 9.

13. Write the equation of a sine curve that has an amplitude of 4 and is shifted 9 units up.

14. Write the equation of a cosine curve that has an amplitude of 6 and is shifted 10 units down.

15. Find the frequency and phase shift of $y = 3 + \frac{1}{2}\sin(5x + 4)$.

16. Find the frequency and phase shift of $y = \cos(6x - 3)$.

17. Find the asymptotes of the function $y = \tan\left(\frac{\pi}{2}x + \frac{\pi}{4}\right)$.

18. Find the asymptotes of the function $y = \tan\left(\frac{\pi}{4}x + \frac{\pi}{8}\right)$.

19. Find the phase shift and amplitude of the function $y = \tan\left(\frac{\pi}{4}x + \frac{\pi}{12}\right)$.

20. As a function approaches infinity, where is its reciprocal?

21. What is always true of the sign of a function and its reciprocal?

Questions 22–24 refer to the function $y = 4\csc\left(\frac{\pi}{2}x + \frac{\pi}{3}\right)$.

22. Find the function's period.

23. Find the function's phase shift.

24. Determine the function's asymptotes.

25. Describe the main difference between the graphs of tangent and cotangent.

Answers and Solutions

1. **Answer:** 1.309   Set up a proportion based upon the fact that $180° = \pi$ radians:

$$\frac{75}{180} = \frac{x}{\pi}$$

$$180x = 75\pi$$

$$x = \frac{75\pi}{180} = \frac{5\pi}{12} = 1.309$$

2. **Answer:** 30°   You are looking to express $\frac{\pi}{6}$ radians, but remember that $180° = \pi$ radians, so

$$\frac{\pi}{6} = \frac{180°}{6} = 30°$$

You also could have solved the proportion: $\dfrac{\frac{\pi}{6}}{\pi} = \dfrac{x}{180}$

$$\pi x = \frac{180\pi}{6} = 30\pi$$

$$x = 30°$$

3. **Answer:** $\frac{7\pi}{6}$ or $\frac{7}{6}\pi$ First the long way: $\dfrac{210°}{180°} = \dfrac{\theta}{\pi}$

$$\frac{\cancel{210}}{\cancel{180}} \frac{7}{6} = \frac{\theta}{\pi}$$

$$6\theta = 7\pi$$

$$\theta = \frac{7\pi}{6} \text{ or } \frac{7}{6}\pi$$

Now the short way: You know that $210° = 180° + 30°$, and you know from the first table in this chapter that $180° = \pi$ and $30° = \frac{\pi}{6}$. Add them:

$$\pi + \frac{\pi}{6} = \frac{6\pi}{6} + \frac{\pi}{6} = \frac{7\pi}{6}$$

4. **Answer:** 7.8 cm   An arc length of one radian is equal to the radius of the circle. This circle's radius is half the diameter, or 6 cm.

$$1.3 \times 6 = 7.8 \text{ cm}$$

5. **Answer:** sine, cosine   Tangent is a ratio of sine to cosine. The three reciprocal functions are constructed by inverting these three.

6. **Answer:** sine   Sine is opposite over hypotenuse, but in the unit circle, hypotenuse is the radius of the circle whose value is 1. That makes sine the height of the side opposite it, or y.

7. **Answer:** $-\frac{3\pi}{2}, -\frac{\pi}{2}, \frac{\pi}{2}, \frac{3\pi}{2}$ Rather than trying to think in radian terms, think in the unit you're more accustomed to, degrees. The sine function reaches $|1|$ four times between -2π and 2π; is positive 1 at 90° and 360° to the left of that at −270°; and is −1 at 270°, and 360° left of that, at −90°.

    Those points have x-coordinates $-\frac{3\pi}{2}, -\frac{\pi}{2}, \frac{\pi}{2}$, and $\frac{3\pi}{2}$. Check out the following figure:

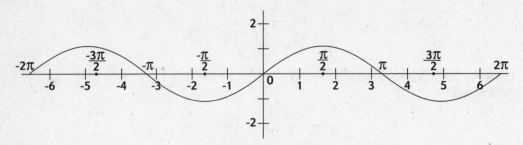

Sine from −2π to 2π.

8. **Answer:** $-2\pi, -\pi, 0, \pi, 2\pi$ Look at the figure directly preceding this answer. Sine hits the x-axis at −360°, −180°, 0°, 180°, and 270°. Those are x-coordinates: −2π, −π, 0, π, and 2π.

9. **Answer:** $\frac{\sqrt{5}}{8}, \frac{\sqrt{5}}{8}$ Sine is a periodic function, whose period is 2π.

    Subtracting any multiple of 2π from it will not change its value, since it returns you to the same point on the circle.

10. **Answer:** $\frac{1}{2}, \frac{1}{2}$ Cosine is a periodic function, whose period is 2π.

    Adding or subtracting any multiple of 2π will not change its value, since it returns you to the same point on the circle.

11. **Answer:** $y = -6 + 11 \sin x$ or $y = 11 \sin x - 6$ The first form of the answer is the preferred one, since it keeps a and b in their proper places. In the equation $y = a + b \sin x$, a tells the number that the curve has been vertically displaced from the axis, and b is the amplitude of the curve. This curve has shifted down 6, so $a = -6$, and its amplitude is 11, so $b = 11$.

12. **Answer:** $y = 7 + 9 \cos x$ In the equation $y = a + b \cos x$, a tells the number that the curve has been vertically displaced from the axis, and b is the amplitude of the curve. This curve has shifted up 7, so $a = +7$, and its amplitude is 9, so $b = 9$.

13. ***Answer:*** $y = 9 + 4 \sin x$ In the equation $y = a + b \sin x$, a tells the number that the curve has been vertically displaced from the axis, and b is the amplitude of the curve. Don't let the equation's statement of those values in reverse order fool you. This curve has shifted up 9, so $a = +9$, and its amplitude is 4, so $b = 4$.

14. ***Answer:*** $y = -10 + 6 \cos x$ In the equation $y = a + b \cos x$, a tells the number that the curve has been vertically displaced from the axis, and b is the amplitude of the curve. Don't let the equation's statement of those values in reverse order fool you. This curve has shifted down 10, so $a = -10$, and its amplitude is 6, so $b = 6$.

15. ***Answer:*** 5, 4   In the equation $y = a + b \sin(cx + d)$, a tells the number that the curve has been vertically displaced from the axis, and b is the amplitude of the curve; c is the frequency, and d is the phase shift. Here, $c = 5$ and $d = 4$.

16. ***Answer:*** 6, −3   In the equation $y = a + b \sin(cx + d)$, a tells the number that the curve has been vertically displaced from the axis, and b is the amplitude of the curve; c is the frequency, and d is the phase shift. Here, $c = 6$ and $d = -3$.

17. ***Answer:*** $\frac{1}{2}, -\frac{3}{2}$ To find asymptotes, solve these for x:

$$cx + d = \frac{\pi}{2} \quad \text{and} \quad cx + d = -\frac{\pi}{2} \text{ for } x.$$

$$\frac{\pi}{2}x + \frac{\pi}{4} = \frac{\pi}{2} \quad \text{and} \quad \frac{\pi}{2}x + \frac{\pi}{4} = -\frac{\pi}{2}$$

$4 \times$ everything clears denominators:

$$4\left(\frac{\pi}{2}x + \frac{\pi}{4} = \frac{\pi}{2}\right) \qquad 4\left(\frac{\pi}{2}x + \frac{\pi}{4} = -\frac{\pi}{2}\right)$$

Multiply: $2\pi x + \pi = 2\pi$ $2\pi x + \pi = -2\pi$

Collect terms: $2\pi x = 2\pi - \pi$ $2\pi x = -2\pi - \pi$

Now add: $2\pi x = \pi$ $2\pi x = -3\pi$

. . . and divide: $x = \frac{1}{2}$ $x = -\frac{3}{2}$

18. ***Answer:*** $\frac{3}{2}$ or $1\frac{1}{2}$, $-\frac{5}{2}$ or $-2\frac{1}{2}$ To find asymptotes, solve these for x:

$$cx + d = \frac{\pi}{2} \quad \text{and} \quad cx + d = -\frac{\pi}{2} \text{ for } x.$$

$$\frac{\pi}{4}x + \frac{\pi}{8} = \frac{\pi}{2} \quad \text{and} \quad \frac{\pi}{4}x + \frac{\pi}{8} = -\frac{\pi}{2}$$

$8 \times$ everything clears denominators:

$$8\left(\frac{\pi}{4}x + \frac{\pi}{8} = \frac{\pi}{2}\right) \qquad 8\left(\frac{\pi}{4}x + \frac{\pi}{8} = -\frac{\pi}{2}\right)$$

Multiply: $2\pi x + \pi = 4\pi$ $2\pi x + \pi = -4\pi$

Collect terms: $2\pi x = 4\pi - \pi$ $2\pi x = -4\pi - \pi$

Now add: $2\pi x = 3\pi$ $2\pi x = -5\pi$

. . . and divide: $x = \frac{3}{2}$ or $1\frac{1}{2}$ $x = -\frac{5}{2}$ or $-2\frac{1}{2}$

19. **Answer:** $\frac{1}{3}$, none   Phase shift is $\frac{d}{c}$, or $d \div c$.

To solve: $d \div c = \frac{\pi}{12} \div \frac{\pi}{4}$

That becomes reciprocal multiplication: $\frac{\pi}{12} \div \frac{\pi}{4} = \frac{\pi}{12} \times \frac{4}{\pi} = \frac{4\pi}{12\pi} = \frac{1}{3}$

Tangent functions have no amplitude.

20. **Answer:** approaching zero   It has been stated that as a function approaches zero, its reciprocal is approaching positive or negative infinity. The converse, then, must also be true.

21. **Answer:** same   The sign of a function is always the same as the sign of its reciprocal function. Cosine is positive in quadrant IV, so secant must also be positive in that quadrant.

22. **Answer:** 4

$$cx = 2\pi$$
$$\frac{\pi}{2}x = 2\pi$$
$$x = \frac{2\pi}{\frac{\pi}{2}} \to \to x = \frac{2\pi}{1} \times \frac{2}{\pi}$$
$$x = 4$$

23. **Answer:** $\frac{3}{2}$ or $1\frac{1}{2}$

$$cx - d = 0$$
$$\frac{\pi}{2}x - \frac{\pi}{3} = 0$$
$$\frac{\pi}{2}x = \frac{\pi}{3}$$
$$x = \frac{\frac{\pi}{2}}{\frac{\pi}{3}} \to \to x = \frac{\pi}{2} \times \frac{3}{\pi}$$
$$x = \frac{3}{2} \text{ or } 1\frac{1}{2}$$

24. **Answer:** $-\frac{2}{3}, \frac{4}{3}$ or $1\frac{1}{3}, -\frac{10}{3}$ or $-3\frac{1}{3}$

$$cx + d = 0 \qquad\qquad cx + d = \pi \qquad\qquad cx + d = 2\pi$$

$$\frac{\pi}{2}x + \frac{\pi}{3} = 0 \qquad\qquad \frac{\pi}{2}x + \frac{\pi}{3} = \pi \qquad\qquad \frac{\pi}{2}x + \frac{\pi}{3} = 2\pi$$

$$\frac{\pi}{2}x = -\frac{\pi}{3} \qquad\qquad \frac{\pi}{2}x = \pi - \frac{\pi}{3} \qquad\qquad \frac{\pi}{2}x = 2\pi - \frac{\pi}{3}$$

$$x = -\frac{\pi}{3} \times \frac{2}{\pi} = -\frac{2}{3} \qquad x = \frac{2\pi}{3} \times \frac{2}{\pi} = \frac{4}{3} \qquad x = -\frac{5\pi}{3} \times \frac{2}{\pi} = -\frac{10}{3}$$

$$x = -\frac{2}{3} \qquad\qquad x = \frac{4}{3} \text{ or } 1\frac{1}{3} \qquad\qquad x = -\frac{10}{3} \text{ or } -3\frac{1}{3}$$

25. **Answer:** They are inverses   Tangent rises as it moves from left to right, disappearing as it approaches its asymptotes, coming back in at negative infinity and then rising again. Cotangent falls as it moves from left to right, disappearing at negative infinity as it approaches its asymptotes, coming back in at infinity and continuing to fall.

Supplemental Chapter Problems

Solve these problems for even more practice applying the skills from this chapter. The Answer section will direct you to where you need to review.

Problems

1. What is the length of an arc subtended by an angle of 2.0 radians in a circle with a diameter of 15 inches?

2. What is the radian measure of a 135° angle?

3. What is the degree measure of an angle of $\frac{\pi}{9}$?

4. What is the radian measure of a 75° angle?

5. How is cotangent determined in circular functions?

6. What circular function = the x-coordinate of any point on the unit circle?

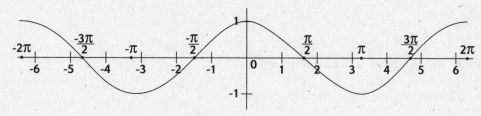

Use the above figure to answer questions 7 and 8.

7. What values of x in the domain of the cosine function between -2π and 2π have a range equal to an absolute value of 1?

8. What values of x in the domain of the cosine function between -2π and 2π have a range equal to 0?

9. If $\cos y = \dfrac{\sqrt{2}}{9}$, what is $\cos (y + 14\pi)$?

10. If $\sin y = \dfrac{3\sqrt{7}}{11}$, what is $\cos (y + 14\pi)$?

11. Write the equation of a cosine curve that is shifted 9 units down and has an amplitude of 6.

12. Write the equation of a sine curve that has an amplitude of 12 and is shifted 7 units up.

13. Write the equation of a cosine curve that is shifted 14 units up and has an amplitude of 3.

14. Write the equation of a sine curve that has an amplitude of 5 and is shifted 8 units down.

15. Find the frequency and phase shift of $y = 4 + \dfrac{1}{2}\cos(3x - 9)$.

16. Find the phase shift and frequency and of $y = \sin(4x + 7)$.

17. Find the asymptotes of the function $y = \tan(3x + 4)$.

18. Find the asymptotes of the function $y = \tan\left(\dfrac{\pi}{4}x + \dfrac{\pi}{12}\right)$.

19. Find the phase shift and amplitude of the function $y = \tan(5x - 20)$.

20. The graph of a function begins at infinity, moves to the left through zero, and disappears toward infinity as it approaches an asymptote. Then it repeats the pattern again and again. Of what function is the graph?

21. What is the sign of the cotangent function in quadrant IV?

22. What is the sign of the cosecant function in quadrant III?

Questions 23–25 refer to the function $y = 3\sec\left(\dfrac{\pi}{6}x + \dfrac{\pi}{6}\right)$.

23. Determine where the function's asymptotes are located.

24. Find the function's phase shift.

25. Find the function's period.

Answers

1. 15 in.   (Understanding Radians, p. 59)

2. $\dfrac{3\pi}{4}$ (Understanding Radians, p. 59)

3. 20°   (Relationships between Degrees and Radians, p. 60)

4. $\dfrac{5\pi}{12}$ (Relationships between Degrees and Radians, p. 60)

5. $\dfrac{\cos\theta}{\sin\theta}$ (The Unit Circle and Circular Functions, p. 62)

6. cosine   (The Unit Circle and Circular Functions, p. 62)

7. $-2\pi, -\pi, 0, \pi, 2\pi$ (Domain versus Range, p. 63)

8. $-\dfrac{3\pi}{2}, -\dfrac{\pi}{2}, \dfrac{\pi}{2}, \dfrac{3\pi}{2}$ (Domain versus Range, p. 63)

9. $\dfrac{\sqrt{2}}{9}$ (Periodic Functions, p. 66)

10. $\dfrac{3\sqrt{7}}{11}$ (Periodic Functions, p. 66)

11. $y = -9 + 6\cos x$ (Vertical Displacement and Amplitude, p. 70)

12. $y = 7 + 12\sin x$ (Vertical Displacement and Amplitude, p. 70)

13. $y = 14 + 3\cos x$ (Vertical Displacement and Amplitude, p. 70)

14. $y = -8 + 5\sin x$ (Vertical Displacement and Amplitude, p. 70)

15. 3, −9   (Frequency and Phase Shift, p. 74)

16. 7, 4   (Frequency and Phase Shift, p. 74)

17. $\dfrac{\pi - 8}{6}, \dfrac{-\pi - 8}{6}$ (Asymptotes, p. 77)

18. $\dfrac{5}{3}, -\dfrac{7}{3}$ (Asymptotes, p. 77)

19. −4, undefined   (Graphing Tangents, p. 77)

20. cotangent   (Cotangent's Graph, p. 77)

21. negative   (Graphing the Reciprocal Functions, p. 78)

22. negative   (Graphing the Reciprocal Functions, p. 78)

23. −1, 5, 11   (Graphing the Reciprocal Functions, p. 78)

24. 1   (Graphing the Reciprocal Functions, p. 78)

25. 12   (Graphing the Reciprocal Functions, p. 78)

Chapter 3
Trigonometry of Triangles

E very triangle is made up of three line segments, which meet to form three angles. The sides of every triangle are related in size to the angles opposite them. In the case of right triangles, a right angle can be included between the shorter two of those sides. If you know two of the sides of a right triangle, you can find the length of the third side by using the Pythagorean theorem. In this chapter, we'll examine how you can find all parts of a triangle if you know at least one side. If you're dealing with a right triangle, you can use the trigonometric ratios to find the missing parts. For a general triangle (non-right) you need different techniques, which we'll discuss later in the chapter.

Finding Missing Parts of Right Triangles

Finding all the missing parts of a triangle using three known parts is called **solving the triangle.** Look at the way in which the angles and sides of the right triangle $\angle ABC$ are labeled in the following figure. Each angle is named by an uppercase letter, and each side is named by a lowercase letter of the angle it is opposite.

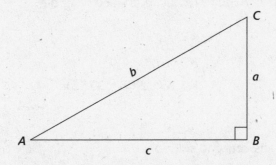

Labeling sides according to angles opposite.

Since the right triangle already has a known angle of 90°, you could solve it if you were given the measures of two of the sides or if you were given the measure of one side and one other angle.

Let's say $\angle A = 30°$ and $a = 12$. You know immediately, since the sum of the angles is 180°, $\angle C$ must be 60° $(30° + 90° + 60° = 180°)$. You can use sin and tan functions to find the remaining two sides of $\angle ABC$, like this:

$$\sin A = \frac{a}{b} \qquad\qquad \tan A = \frac{a}{c}$$
$$\sin 30° = \frac{12}{b} \qquad\qquad \tan 30° = \frac{12}{c}$$
$$b \sin 30° = 12 \qquad\qquad c \tan 30° = 12$$
$$b = \frac{12}{\sin 30°} \qquad\qquad c = \frac{12}{\tan 30°}$$
$$b = \frac{12}{0.5} \qquad\qquad c = \frac{12}{0.577}$$
$$b = 24 \qquad\qquad c = 20.80$$

The sides given previously might also have been found using the relations cosecant and cotangent, since:

$$\csc A = \frac{b}{a} \quad \text{and} \quad \cot A = \frac{c}{a}$$

You might even have found this solution in an easier way, since no division is involved.

Solve the triangle in the preceding figure, this time using 16 for a and 26 for c.

First, $\tan A = \frac{16}{26} = \frac{8}{13} = 0.615$.

Using a scientific calculator or the tables on p. 297, $m\angle A = 31.6°$.

That means $m\angle C = 90° - 31.6° = 58.4°$. Finally, you need to find b:

$$\csc A = \frac{b}{16}$$
$$b = 16(\csc A)$$
$$b = 16(1.908 = 30.53)$$

Example Problems
These problems show the answers and solutions.

1. In right triangle $\triangle PQR$, $m\angle P = 40°$ and hypotenuse $q = 12$. Solve the triangle.

    answer: $m\angle Q = 90°$, $m\angle R = 50°$, $p = 7.72$, $r = 9.19$

    You know that the right angle is $\angle Q$ since q is the hypotenuse. $90° + 40° = 130°$, so $180° - 130° = 50°$. You can find the sides by means of sine, and/or cosine, and/or secant, and/or cosecant. The easiest way is probably as follows:

    $$\sin R = \frac{r}{q} \qquad\qquad \sin P = \frac{p}{q}$$
    $$r = q \sin R \qquad\qquad p = q \sin P$$
    $$r = 12 \sin 50° \qquad\qquad p = 12 \sin 40°$$
    $$r = 12(0.766) \qquad\qquad p = 12(0.643)$$
    $$r = 9.19 \qquad\qquad p = 7.72$$

2. In right triangle $\angle DEF$, $m\angle D = 60°$ and hypotenuse $e = 12$. Solve the triangle.

    answer: $m\angle E = 90°$, $m\angle F = 30°$, $d = 6\sqrt{3}$, and $f = 6$

    You know that the right angle is $\angle E$ since e is the hypotenuse. $90° + 60° = 150°$, so $m\angle F = 180° - 150° = 30°$. You can find the sides by means of sine, and/or cosine, and/or secant, and/or cosecant. The easiest, however, is to remember from geometry that in a 30-60-90 right triangle, the side opposite the 30° angle is half the hypotenuse, and the side opposite the 60° angle is half the hypotenuse times $\sqrt{3}$.

3. In right triangle $\angle ABC$, $m\angle A = 25°$ and C is the right angle. $b = 10$. Solve the triangle.

    answer: $m\angle B = 65°$, $a = 4.66$, $c = 11.04$

    $90° + 25° = 115°$, so $m\angle B = 180° - 115° = 65°$. You can solve this in many ways, such as this one:

$$\tan A = \frac{a}{b} \qquad\qquad \sin B = \frac{b}{c}$$

$$a = b\tan A \qquad\qquad\; b = c\sin B$$

$$a = 10\tan 25° \qquad\quad c = \frac{b}{\sin B}$$

$$a = 10(0.466) \qquad\quad c = \frac{10}{\sin 65°}$$

$$a = 4.66 \qquad\qquad\quad\; c = \frac{10}{0.906}$$

$$c = 11.04$$

Angles of Elevation and Depression

Certain angles have special names depending on the application that is being performed. When looking up at the top of a tree, the angle formed by your line of sight and the ground is known as an **angle of elevation.** The following figure shows such an angle. For purposes of computation, unless otherwise noted, it is presumed that the person's eye is at ground level.

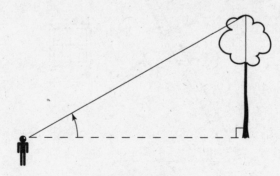

An angle of elevation.

When a pilot looks down from an airplane's cockpit, the angle formed between his line of sight and the ground is known as an **angle of depression.** Such an angle is shown in the following figure.

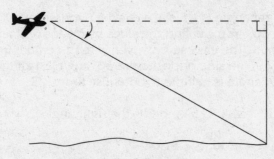

An angle of depression.

Angles of elevation and depression are useful in solving many different types of practical, real-life problems through the use of trigonometry. The Example Problems that follow will give you a few such instances.

Example Problems

These problems show the answers and solutions.

1. The foot of a ladder is 6 feet away from the bottom of a wall. The ladder forms a 40° angle of elevation with the ground. How long is the ladder?

   answer: **7.83 ft.**

   The first thing you should do is to draw a picture like the one that follows.

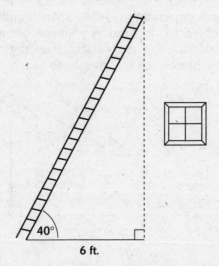

40°

6 ft.

Since the sides involved are the adjacent and the hypotenuse, solve by using cosine.

$$\cos 40° = \frac{6}{x}$$
$$(\cos 40°)(x) = 6$$
$$x = \frac{6}{\cos 40°}$$
$$x = \frac{6}{0.766}$$
$$x = 7.83 \text{ ft.}$$

2. The pilot of an airplane flying at 26,000 feet spots another plane flying at 24,000 feet. The angle of depression to the second plane is 45°. What is the length of the pilot's line of sight to the second plane?

answer: 2,828.85 ft.

Here again, drawing a picture like the one that follows will help. Note the line of sight (hypotenuse) is what you're looking for. The side of the relevant triangle is 2,000 ft (26,000 − 24,000 = 2,000).

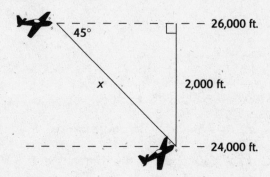

$$\sin 45° = \frac{2000}{x}$$
$$(\sin 45°)(x) = 2000$$
$$x = \frac{2000}{\sin 45°}$$
$$x = \frac{2000}{0.707}$$
$$x = 2828.85 \text{ ft.}$$

3. An observer is standing 16 m from the bottom of a tree. Assume that her eye is at ground level. The angle of elevation from the observer to the top of a tree is 60°. How tall is the tree?

answer: 27.71 m

First, picture the problem as in the following image.

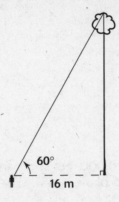

60°

16 m

Here, you know the adjacent side of the triangle (with respect to the angle of elevation) and want to find the opposite side, so use tangent:

$$\tan 60° = \frac{x}{16}$$
$$x = 16(\tan 60°)$$
$$x = 16(1.732)$$
$$x = 27.71 \text{ m}$$

Work Problems

Use these problems to give yourself additional practice.

1. In right triangle $\triangle PQR$, $m\angle P = 26°$ and hypotenuse $q = 16$. Solve the triangle.

2. In right triangle $\triangle ABC$, $m\angle A = 40°$ and C is the right angle. $b = 20$. Solve the triangle.

3. The top of a ladder is leaning against a lamppost. Its bottom is 4 feet away from the base of the pole. The ladder forms a 75° angle of elevation with the ground.

   a. How long is the ladder?

   b. How far up the pole is the top of the ladder?

4. Richard stood on top of a mountain on a cloudless day. He looked out toward the valley below at an angle of depression of 50° through an optical rangefinder. The rangefinder showed the distance to the valley floor as being 8,000 meters. How high above the valley floor was Richard standing? The rangefinder he was looking through was 5 feet above the ground.

5. Hailee is standing 14 feet from the bottom of a tree. Assume her eye is 4.5 feet above ground level. The angle of elevation from her eye to the top of a tree is 75°. How tall is the tree?

Worked Solutions

1. **$m\angle Q = 90°$, $m\angle R = 64°$, $p = 7.01$, $r = 14.38$**   Drawing a right triangle would help to visualize the problem. First, figure out which angle is the right angle. You know that the right angle is $\angle Q$ since q is the hypotenuse. $90° + 26° = 116°$, so $m\angle R = 180° - 116° = 64°$. You can find the sides by means of sine, and/or cosine, and/or secant, and/or cosecant. The simplest way is probably as follows:

$$\sin R = \frac{r}{q} \qquad\qquad \sin P = \frac{p}{q}$$
$$r = q\sin R \qquad\qquad p = q\sin P$$
$$r = 16\sin 64° \qquad\quad p = 16\sin 26°$$
$$r = 16(0.899) \qquad\quad p = 16(0.438)$$
$$r = 14.384 \qquad\qquad p = 7.008$$

2. **$m\angle B = 50°$, $a = 16.78$, $c = 26.11$**   Drawing a diagram will make it easier. $90° + 40° = 130°$, so $m\angle B = 180° - 130° = 50°$. You can solve this triangle in many ways, and the following method is just one of them:

$$\tan A = \frac{a}{b} \qquad\qquad \sin B = \frac{b}{c}$$
$$a = b\tan A \qquad\qquad b = c\sin B$$
$$a = 20\tan 40° \qquad\qquad c = \frac{b}{\sin B}$$
$$a = 20(0.839) \qquad\qquad c = \frac{20}{\sin 50°}$$
$$a = 16.78 \qquad\qquad c = \frac{20}{0.766}$$
$$c = 26.11$$

3. **a. 15.44 ft.; b. 14.92 ft.**   The first thing you should do is to draw a picture like the following one. It's not a very complicated diagram, since you'll end up with a right triangle.

   Since the sides involved are the adjacent and the hypotenuse, solve by using cosine.

$$\cos 75° = \frac{4}{x}$$
$$(\cos 75°)(x) = 4$$
$$\sin 75° = \frac{y}{15.44}$$
$$x = \frac{4}{\cos 75°}$$
$$y = 15.44(\sin 75°)$$
$$x = \frac{4}{0.259}$$
$$y = 15.44(0.966)$$
$$\text{a)}\, x = 15.44 \text{ ft.}$$
$$\text{b)}\, y = 14.92 \text{ ft.}$$

4. **6126.4 m**   If you can do this one without a diagram, you're better than the author! Note Richard's line of sight is the hypotenuse. His height above the valley is the side opposite the angle of depression.

$$\sin 50° = \frac{x}{8000}$$
$$x = (8000)(\sin 50°)$$
$$x = (8000)(0.766)$$
$$x = 6128 \text{ m}$$

But his eye (on the range finder) is 5 feet above the ground. That's about 1.6 m, so he is actually standing about 6126.4 m above the valley's floor.

5. **56.75 ft.**   First, picture the problem. Here, you know the adjacent side of the triangle (with respect to the angle of elevation) and want to find the opposite side, so use tangent:

$$\tan 75° = \frac{x}{14}$$
$$x = 14(\tan 75°)$$
$$x = 14(3.732)$$
$$x = 52.25 \text{ ft.}$$

But wait. That's the distance from Hailee's eye to the tree top. To find the height, you must add the distance from her eye to the ground. 52.25 + 4.5 = 56.75 ft.

The Law of Sines

The preceding two sections deal with solving right triangles, but formulas have been developed for solving all triangles, whether the triangles are right or not. The first formula you'll look at is known as the **Law of Sines.** It says that if you know the measures of any two angles of a triangle and the length of the side opposite one of those angles, you can find the side opposite the second angle. Alternately, if you know the lengths of any two sides of a triangle and the measure of an angle opposite one of those sides, you can find the measure of the angle opposite the second side. It looks like this:

$$\frac{\sin A}{a} = \frac{\sin B}{b} = \frac{\sin C}{c} \quad \text{or} \quad \frac{a}{\sin A} = \frac{b}{\sin B} = \frac{c}{\sin C}$$

These relationships may be stated as follows: In any triangle, the ratio of the sine of an angle to the length of that angle's opposite side is constant, or the ratio of the length of any side to the sine of the angle opposite that side is constant. Since the sine function is positive in both the first and second quadrants, the Law of Sines is valid for obtuse triangles as well as acute ones.

Example Problems

These problems show the answers and solutions.

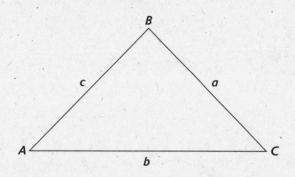

1. Solve the triangle in the preceding figure if $m\angle A = 50°$, $m\angle C = 60°$, and $b = 16$.

    answer: $m\angle B = 70°$, a = 13.04, c = 14.74

    Since the total sum of the angles $= 180°$, and $50° + 60° = 110°$, $m\angle B = 180° - 110° = 70°$

    From the Law of Sines you get: $\dfrac{a}{\sin A} = \dfrac{b}{\sin B} = \dfrac{c}{\sin C}$

    That means $\dfrac{a}{\sin 50°} = \dfrac{16}{\sin 70°} = \dfrac{c}{\sin 60°}$

    To solve, make two proportions:

    $\dfrac{a}{\sin 50°} = \dfrac{16}{\sin 70°}$ and $\dfrac{16}{\sin 70°} = \dfrac{c}{\sin 60°}$

    Solve each:
    $$\dfrac{a}{0.766} = \dfrac{16}{0.940} \qquad\qquad \dfrac{16}{0.940} = \dfrac{c}{0.866}$$
    $$0.940a = 16(0.766) \qquad\qquad 0.940c = 16(0.866)$$
    $$0.940a = 12.256 \qquad\qquad 0.940c = 13.856$$
    $$a = 13.04 \qquad\qquad\qquad c = 14.74$$

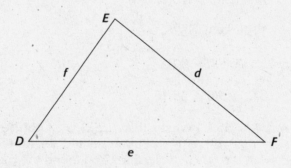

2. Referring to the triangle in the preceding figure, if $m\angle D = 75°$, $m\angle E = 65°$, and $d = 12$ cm, find f.

    answer: 7.99 cm

    You'll use the Law of Sines: $\dfrac{d}{\sin D} = \dfrac{f}{\sin F}$

    Of course, you'll need $m\angle F$. $m\angle F = 180° - (65° + 75°)$

$$180° - 140° = 40°$$

$$\frac{12}{\sin 75°} = \frac{f}{\sin 40°}$$

$$\frac{12}{0.966} = \frac{f}{0.643}$$

$$0.966f = (0.643)(12)$$

$$0.966f = 7.716$$

$$f = 7.99$$

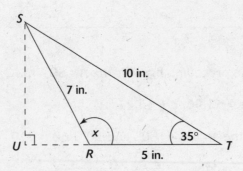

3. Using the Law of Sines and △RST in the above figure, find $m\angle SRT$.

answer: 124.9°

First, set up the equation: $\dfrac{r}{\sin SRT} = \dfrac{t}{\sin T}$

Use $\angle SRT$ because it's the angle you are looking for and $\angle T$ because it's the only angle whose measure you know. Of course, you know all three sides.

$$\frac{10}{\sin SRT} = \frac{7}{\sin 35}$$

Cross multiply: $7 (\sin SRT) = 10\sin 35°$

Find $\sin 35°$: $7 (\sin SRT) = 10(0.574)$

Multiply: $7 (\sin SRT) = 5.74$

Divide both sides by 7: $\sin SRT = 0.82$

Find arcsin (or $\sin^{-1}$) 0.82: $\angle SRT = 55.1°$

There you have it: $m\angle SRT = 55.1°$. What's that? $\angle SRT$ is an obtuse angle? Well, looking at the figure, it's apparent that you're right. But arcsin and sin, for that matter, only exist for angles up to 90°. So what did you actually find? You found the measure of $\angle SRU$, the quadrant II reference angle for $\angle SRT$. To get $m\angle SRT$, subtract $\angle SRU$ from 180°. (You did this in Chapter 1.) $180° - 55.1° = 124.9°$, which is the answer.

Work Problems

Use these problems to give yourself additional practice.

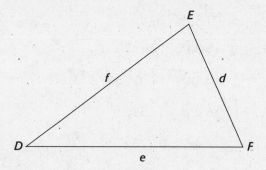

Problems 1 and 2 refer to $\angle DEF$ in the preceding figure.

1. If $m\angle D = 55°$, $m\angle E = 45°$, and $d = 15$ cm, find the length of f.

2. If $m\angle E = 75°$, $e = 15$ cm, and $f = 12$ cm, find $m\angle F$.

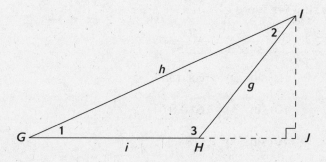

Problems 3 and 4 refer to $\triangle GHI$ in the preceding figure.

3. Using the Law of Sines and $\triangle GHI$ find $m\angle 3$ if $m\angle G = 35°$, $g = 10$ in., and $h = 16$ in.

4. Using the information found in Problem 3, find the length of i.

Worked Solutions

1. **18.04 cm**   Use the Law of Sines: $\dfrac{d}{\sin D} = \dfrac{f}{\sin F}$

    Of course you'll need $m\angle F$: $m\angle F = 180° - (55° + 45°)$

    $180° - 100° = 80°$

    $$\frac{15}{\sin 55°} = \frac{f}{\sin 80°}$$

    $$\frac{15}{0.819} = \frac{f}{0.985}$$

    $$0.819f = (0.985)(15)$$

    $$0.819f = 14.775$$

    $$f = 18.04$$

2. **50.6°**

First, state the Law of Sines as $\dfrac{e}{\sin E} = \dfrac{f}{\sin F}$

Substitute: $\dfrac{15}{\sin 75°} = \dfrac{12}{\sin F}$

Cross multiply: $15\sin F = 12\sin 75°$

Look up sin sin 75°: $15\sin F = 12(0.966)$

Multiply: $15\sin F = 11.592$

Divide both sides by 15: $\sin F = 0.773$

Find arcsin (or $\sin^{-1}$) of 0.773: $F = 50.6°$

3. **113.3°**

First, set up the equation: $\dfrac{h}{\sin\angle 3} = \dfrac{g}{\sin G}$

We used $\angle 3$ because there are two angles at H, and we used $\angle G$ because it's the only angle whose measure we were given.

$$\frac{16}{\sin\angle 3} = \frac{10}{\sin 35°}$$

Cross multiply: $10(\sin\angle 3) = 16\sin 35°$

Find sin 35°: $10(\sin\angle 3) = 16(0.574)$

Multiply: $10(\sin\angle 3) = 9.184$

Divide both sides by 10: $\sin\angle 3 = 0.9184$

Find arcsin 0.9184: $\angle 3 = 66.7°$

There you have it: $m\angle 3 = 66.7°$. But $\angle 3$ is an obtuse angle and arcsin and sin only exist for angles up to 90°. So what did you actually find? You found the measure of $\angle IHJ$, what would have been the quadrant II reference angle for $\angle 3$, had that angle been drawn in standard position. To get $m\angle 3$, subtract $\angle IHJ$ from 180°. (You saw this in Chapter 1.) $180° - 66.7° = 113.3°$, which is the answer.

4. **9.15 in.**   Use the Law of Sines, $\dfrac{i}{\sin\angle 2} = \dfrac{g}{\sin G}$

Use $\angle 2$ because there are two angles at I and $\angle G$ because it's the angle whose measure you were given at the start.

Of course you'll need $m\angle F$. $m\angle F = 180° - (35° + 113.3°)$

$180° - 148.3° = 31.7°$

If you have a scientific calculator, you can enter that value. If you're using the table on p. 297, you'll have to interpolate.

Substitute known values: $\dfrac{i}{\sin 31.7°} = \dfrac{10}{\sin 35°}$

Look up the sines: $\dfrac{i}{0.525} = \dfrac{10}{0.574}$

Cross multiply: $0.574i = (0.525)(10)$

Clear parentheses: $0.574i = 5.25$

Divide both sides by 0.574: $i = 9.15$

The Law of Cosines

It was fortunate that the Law of Sines was an easy formula to memorize. Every bit as useful as the Law of Sines is the **Law of Cosines,** but, unfortunately, it is much less easy to memorize. Like the Law of Sines, the Law of Cosines is true for any triangle: acute, right, or obtuse. Since the cosine function is negative in quadrant II, a cosine < 0 indicates that an angle is obtuse. That is, were the angle drawn in standard position, it would terminate in quadrant II. The Law of Cosines is derived from the distance formula, which, you may recall, was derived from the Pythagorean theorem. It may be written in any one of three ways, all of which are shown here. The statements of the law presume $\triangle ABC$ with sides a, b, and c in the usual places, and it goes like this:

$$a^2 = b^2 + c^2 - 2bc\cos A$$
$$b^2 = a^2 + c^2 - 2ac\cos B$$
$$c^2 = a^2 + b^2 - 2ab\cos C$$

Look at the last equation. Interestingly, if the angle in question happens to be a 90° angle, cosine of 90° = 0, and the entire formula turns into the Pythagorean theorem.

Notice that each statement of the Law of Cosines starts with the side opposite the cosine of the angle that ends that equation. On the other side of the equal sign is the sum of the squares of the other two sides, minus 2 times the sides times the cosine previously mentioned.

If you had an angle to work with, the Law of Sines is much easier to use than the Law of Cosines. The Law of Sines requires the measure of at least one angle to be known. When no angle is known, but all sides are, it's time to reach for the Law of Cosines. Since it is almost always used for finding an angle, the three forms of the Law of Cosines shown previously may be rewritten as follows:

$$\cos B = \frac{a^2 + c^2 - b^2}{2ac}$$

$$\cos A = \frac{b^2 + c^2 - a^2}{2bc}$$

$$\cos C = \frac{a^2 + b^2 - c^2}{2ab}$$

Next, you'll see how the Law of Cosines is used.

Example Problems

These problems show the answers and solutions.

1. In $\triangle ABC$, $b = 10$, $c = 8$, and $\cos A = \frac{1}{8}$. Find a.

    ***answer:* 12**

    Use the $\cos A$ = form: $\qquad \cos A = \dfrac{b^2 + c^2 - a^2}{2bc}$

    Substitute: $\qquad\qquad\quad \dfrac{1}{8} = \dfrac{10^2 + 8^2 - a^2}{2 \cdot 10 \cdot 8}$

    Multiply: $\qquad\qquad\quad\;\; \dfrac{1}{8} = \dfrac{100 + 64 - a^2}{2 \cdot 80}$

    Combine like terms: $\qquad \dfrac{1}{8} = \dfrac{164 - a^2}{160}$

    Cross multiply: $\qquad 8(164 - a^2) = 160$

    Divde both sides by 8: $\quad 164 - a^2 = 20$

    Collect terms: $\qquad\qquad -a^2 = 20 - 164$

    Add and multiply by -1: $\qquad a^2 = 144$

    $\sqrt{}$ both sides: $\qquad\qquad a = 12$

2. In $\triangle ABC$, $a = 10$, $b = 14$, and $c = 12$. Find $\angle B$.

    ***answer:* 78.5°**

    Let's use the $\cos B$ = form: $\quad \cos B = \dfrac{a^2 + c^2 - b^2}{2ac}$

    Substitute: $\qquad\qquad\quad \cos B = \dfrac{10^2 + 12^2 - 14^2}{2 \cdot 10 \cdot 12}$

    Multiply: $\qquad\qquad\quad\;\; \cos B = \dfrac{100 + 144 - 196}{2 \cdot 120}$

    Combine like terms: $\qquad \cos B = \dfrac{48}{240} = \dfrac{1}{5} = 0.200$

    Find arccos (or $\cos^{-1}$) 0.200: $\quad \angle B = 78.5°$

3. In $\triangle PQR$, $p = 8$, $q = 10$, and $r = 13$. Find $\angle$s P, Q, and R.

    ***answer:* $\angle P$ = 38°, $\angle Q$ = 50.4°, and $\angle R$ = 91.6°**

    Let's use the $\cos P$ = form: $\quad \cos P = \dfrac{q^2 + r^2 - p^2}{2qr}$

    Substitute: $\qquad\qquad\quad \cos P = \dfrac{10^2 + 13^2 - 8^2}{2 \cdot 10 \cdot 13}$

    Multiply: $\qquad\qquad\quad\;\; \cos P = \dfrac{100 + 169 - 64}{2 \cdot 130}$

    Combine like terms: $\qquad \cos P = \dfrac{205}{260} = \dfrac{41}{52} = 0.788$

    Find arccos (or $\cos^{-1}$) 0.788: $\quad \angle P = 38°$

While $\angle P$ did not exactly make 38°, it was very very close. Next, you can find either $\angle Q$ or $\angle R$.

Opt for $\angle Q$ because 10 is an easier number to deal with than 13. Switch to the Law of Sines, since you now have an angle.

$$\frac{p}{\sin P} = \frac{q}{\sin Q}$$

Substitute: $\dfrac{8}{\sin 38°} = \dfrac{10}{\sin Q}$

Cross multiply: $8\sin Q = 10\sin 38°$

Find sin 52°: $8\sin Q = 10 \cdot 0.616$

Multiply: $8\sin Q = 6.16$

Divide both sides by 8: $\sin Q = 0.77$

Find arcsin (or $\sin^{-1}$) 0.77: $\angle Q = 50.4°$

So $\angle R = 180 - (38° + 50.4°) = 180° - 88.4° = 91.6°$

Work Problems

Use these problems to give yourself additional practice.

1. In $\triangle ABC$, $a = 10$, $b = 18$, and $\cos C = \dfrac{7}{15}$. Find c.

2. In $\triangle ABC$, $a = 8$, $b = 13$, and $c = 15$. Find $\angle B$.

3. In $\triangle ABC$, $a = 7$, $b = 8$, and $c = 5$. Find $\angle C$.

4. In $\triangle ABC$, $a = 7$, $b = 5$, and $c = 3$. Find $\angle A$.

Worked Solutions

1. **16**   Use the $\cos C =$ form: $\cos C = \dfrac{a^2 + b^2 - c^2}{2ab}$

    Substitute: $\dfrac{7}{15} = \dfrac{10^2 + 18^2 - c^2}{2 \cdot 10 \cdot 18}$

    Multiply: $\dfrac{7}{15} = \dfrac{100 + 324 - c^2}{2 \cdot 180}$

    Combine like terms: $\dfrac{7}{15} = \dfrac{424 - c^2}{360}$

    Cross multiply: $15(424 - c^2) = 360(7)$

    Divde both sides by 15: $(424 - c^2) = 24(7) = 168$

    Collect terms: $-c^2 = 168 - 424$

    Add and multiply by −1: $c^2 = 256$

    $\sqrt{}$ both sides: $c = 16$

2. **88.5°**

Use the $\cos B =$ form: $\cos B = \dfrac{a^2 + c^2 - b^2}{2ac}$

Substitute: $\cos B = \dfrac{8 + 13^2 - 15^2}{2 \cdot 10 \cdot 15}$

Multiply: $\cos B = \dfrac{64 + 169 - 225}{2 \cdot 150}$

Combine like terms: $\cos B = \dfrac{8}{300} = \dfrac{2}{75} = 0.267$

Find arccos 0.027: $\angle B = 88.5°$

3. **38.2°**

Use the $\cos C =$ form: $\cos C = \dfrac{a^2 + b^2 - c^2}{2ab}$

Substitute: $\cos C = \dfrac{7^2 + 8^2 - 5^2}{2 \cdot 7 \cdot 8}$

Multiply: $\cos C = \dfrac{49 + 64 - 25}{2 \cdot 56}$

Combine like terms: $\cos C = \dfrac{88}{112} = \dfrac{11}{14} = 0.786$

Find arccos 0.786: $\angle C = 38.2°$

4. **120°**

Use the $\cos A =$ form: $\cos A = \dfrac{b^2 + c^2 - a^2}{2bc}$

Substitute: $\cos A = \dfrac{5^2 + 3^2 - 7^2}{2 \cdot 5 \cdot 3}$

Multiply: $\cos A = \dfrac{25 + 9 - 49}{2 \cdot 15}$

Combine like terms: $\cos A = \dfrac{-15}{30} = -\dfrac{1}{2} = -.5$

Find arccos −0.500

Did you find it? If you tried on a calculator, you probably got a "syntax error," and if you tried looking it up in the table, your luck was even worse. You need to look up arccos 0.500, which gives you $\angle C = 60°$, but you're not done. That negative value for cosine tells you that this angle is in quadrant II, so it's an obtuse angle obtained by subtracting the angle you found from 180°: $180° - 60° = 120°$.

Solving General Triangles

The procedure for solving oblique (nonright) triangles can be broken down into several strategies depending upon what is known about the triangle to be solved. In general, the types of triangles are broken down into the same groups that you use to categorize methods for proving triangles congruent. It is assumed here that all parts of the triangle need to be found. If only some of the parts need to be found, the strategy might need to be modified.

SSS

If three sides of a triangle are known, first use the Law of Cosines to find the largest angle (the one opposite the longest side). This is the best way to tell whether the triangle is obtuse, since cosine is negative in the second quadrant. If the cosine is negative, subtract its arcsine from 180° to find the obtuse angle. If the cosine is 0, then the angle is a right angle. When the largest angle is known, the two remaining angles must be acute; although the second angle could be found by using the Law of Cosines, using the Law of Sines is less complicated. When you have two angles, add them together and subtract their sum from 180° to get the third angle.

SAS

When two sides and the angle formed by those two sides are known, use the Law of Cosines to find the third side of the triangle. When that side has been found, find the smaller of the two remaining angles, using the Law of Sines and the shortest of the three sides. If two sides are the same length, you may skip this step, since the two remaining angles may be found using the rule for isosceles triangles, which says sides of equal length are opposite angles of equal measure. If the included angle is not opposite one of the equal sides, subtract it from 180° and divide by 2. If the included angle is opposite one of the equal sides, the angle opposite the other is the same. Finally, add the two angles and subtract from 180° to find the measure of the third angle.

Example Problems

These problems show the answers and solutions.

1. In $\triangle RST$, $r = 86$, $s = 41$, and $t = 62$. Find the angles to the nearest whole degree.

    answer: $m\angle R \approx 112°$, $m\angle S \approx 26°$, $m\angle T \approx 42°$

    The largest angle is $\angle R$: $\cos R = \dfrac{s^2 + t^2 - r^2}{2st}$

    Substitute: $\cos R = \dfrac{41^2 + 62^2 - 86^2}{2(41)(62)}$

    Multiply: $\cos R = \dfrac{1681 + 3844 - 7396}{2(2542)}$

    Combine like terms: $\cos R = \dfrac{-1871}{5084} = -0.371$

    Find arccos −0.371

    Did you notice that the cosine is negative? You need to look up arccos 0.371, which gives you $\angle R = 68.22°$. That negative value for cosine tells you that this angle is in quadrant II, so it's an obtuse angle obtained by subtracting the angle you found from 180°. $180° - 68.22° = 111.78° \approx 112°$.

    On to the Law of Sines: $\dfrac{r}{\sin R} = \dfrac{s}{\sin S}$

    Substitute: $\dfrac{86}{\sin 112°} = \dfrac{41}{\sin S}$

    Use 68° for 112°: $\dfrac{86}{\sin 68°} = \dfrac{41}{\sin S} \rightarrow \dfrac{86}{0.927} = \dfrac{41}{\sin S}$

Cross multiply: $86\sin S = 41(0.927)$

Multiply: $86\sin S = 38.007$

Divide both sides by 86: $\sin S = 0.442$

Find arcsin 0.442:   26.23°, so $m\angle S \approx 26°$

$$m\angle T \approx 180° - (112° + 26°) \approx 42°$$

2. In $\triangle DEF$, $e = 15$, $f = 20$, and $\angle D = 70°$. Solve the triangle rounding all lengths and angles to the nearest whole unit.

   answer: $d \approx 20$, $m\angle E \approx 45°$, $m\angle F \approx 65°$

   You need to find d first: $\cos D = \dfrac{e^2 + f^2 - d^2}{2ef}$

   Substitute: $\dfrac{0.342}{1} = \dfrac{15^2 + 20^2 - d^2}{2(15)(20)}$

   Multiply: $\dfrac{0.342}{1} = \dfrac{225 + 400 - d^2}{2(300)}$

   Combine like terms: $\dfrac{0.342}{1} = \dfrac{625 - d^2}{600}$

   Cross multiply: $(625 - d^2) = (0.342)(600)$

   Combine like terms: $-d^2 = 205.2 - 625 = -419.8$

   Take the square root of both sides: $d \approx 20$

   Now use the Law of Sines to find e: $\dfrac{e}{\sin E} = \dfrac{d}{\sin D}$

   Substitute: $\dfrac{15}{\sin E} = \dfrac{20}{\sin 70°}$

   Find $\sin 70°$: $\dfrac{15}{\sin E} = \dfrac{20}{0.94}$

   Cross multiply: $20\sin E = 15(0.94)$

   Multiply: $20\sin E = 14.1$

   Divide both sides by 20: $20\sin E = 0.705$

   Find arcsin 0.705:   44.8°, so $m\angle E \approx 45°$

   $$m\angle F \approx 180° - (70° + 45°) \approx 65°$$

3. In $\triangle ABC$, $a = 9$, $b = 9$, and $C = 70°$. Solve the triangle rounding all lengths and angles to the nearest whole units.

   answer: $m\angle A = 55°$, $m\angle B = 55°$, $c \approx 10$

   Since legs a and b are equal in length, $\triangle ABC$ is an isosceles triangle, so both angles A and B are equal to half the difference of $180° - 70°$. $180° - 70° = 110°$, and half that is $55°$.

To find side c, use the Law of Sines, $\frac{c}{\sin C} = \frac{a}{\sin A}$

Substitute: $\frac{c}{\sin 70°} = \frac{9}{\sin 55°}$

Cross multiply: $c \sin 55° = 9 \sin 70°$

Look up the sines: $c(0.819) = 9(0.94)$

Multiply: $c(0.819) = 8.46$

Divide both sides by .819: $c = 10.32 \approx 10$

ASA

If you know the degree measures of two of the angles of a triangle and the length of the side between them (also known as the included side), add those angles together and subtract from 180° to get the size of the third angle, if needed. Either the Law of Cosines or the Law of Sines may be used to find the lengths of the other two sides, but why use a sledgehammer to open a walnut? The Law of Sines will do the job just fine, and with a lot less mess.

SAA

Side-Angle-Angle (or as some refer to it, Angle-Angle-Side) is a condition in which two angles of a triangle are known, as well as the length of a side other than the one included between the two angles. Adding those angles and subtracting the sum from 180° will give you the degree measure of the missing angle. The side you were given is opposite one of the two given angles, so pair them up and use the Law of Sines to find the remaining sides of the triangle.

SSA, The Ambiguous Case

If two sides of a triangle are known, as well as an angle not between the two sides, it is not possible to be sure of what the triangle looks like, or whether there even is a triangle. There are six possible solutions as to what the figure looks like. That's why SSA is known as **the ambiguous case.** Suppose that you are given $\angle A$ and sides b and a. See the following figure.

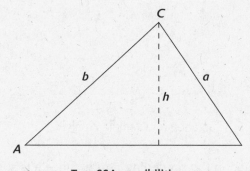

Two SSA possibilities.

Suppose that side a is shorter than the length of h. If so, there is no triangle at all. On the other hand, if a is exactly equal to h, there is one right triangle possible.

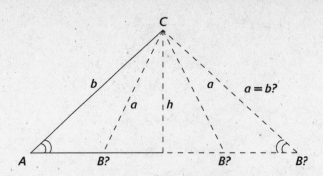

Three more SSA possibilities.

Now suppose that a is longer than h as in the preceding figure. Two triangles are now possible, one acute and one obtuse. If $a = b$, then there is an isosceles triangle formed, and if $a > b$, there is only one triangle possible. These all assume $\angle A$ to be an acute angle. Suppose that $\angle A$ is a right angle or an obtuse angle. Then,

- ❑ $a > b$; there is one triangle.
- ❑ $a = b$; there is one triangle.
- ❑ $a < b$; there is one triangle.

Draw your own diagrams to show that these are the case. Remember that a careful drawing of the circumstances can be used to solve or at least demonstrate most trigonometry problems.

Example Problems

These problems show the answers and solutions.

1. Find the number of triangles that can be formed if $\angle A = 120°$, $b = 16$, and $a = 12$.

    answer: **None**

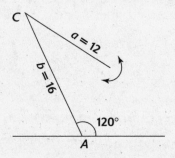

Solution for Example Problem 1.

Notice the representation in the preceding figure. Since a is shorter than b, there can be no triangle.

2. Find the number of triangles that can be formed if $\angle A = 90°$, $b = 25$, and $a = 40$.

    answer: **One**

Since $\angle A = 90°$ and the side opposite it is longer than the side adjacent to it, exactly one triangle can be formed, as shown in the following figure.

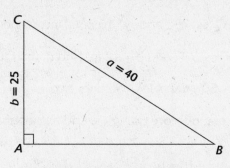

Solution for Example Problem 2.

3. A microwave tower stands atop a hill, which is sloped 26° from the horizontal. An observer, lying on the ground 95 feet down the hill sights the top of that tower at an angle of elevation of 38°. Assuming that observation to be at eye level, how tall—to the nearest foot—is the tower?

answer: **134 ft.**

Drawing the diagram that follows will help you to see that this is an ASA problem, with the angle formed between the base of the tower and the observer 106° $(90° + 26° = 116°)$. That makes $\angle Z = 180° - (116° + 38°) = 180° - 154° = 26°$.

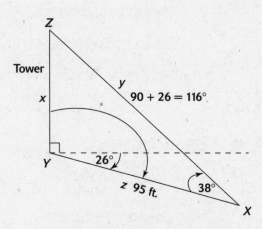

Solution for Example Problem 3.

Find the pole's height by The Law of Sines: $\dfrac{x}{\sin X} = \dfrac{z}{\sin Z}$

Substitute: $\dfrac{x}{\sin 38°} = \dfrac{95}{\sin 26°}$

Cross multiply: $x \sin 26° = 95 \sin 38°$

Look up the sines: $x(0.438) = 95(0.616)$

Multiply: $x(0.438) = 58.52$

Divide both sides by 0.438: $x = 133.6 \approx 134$ ft.

Work Problems

Use these problems to give yourself additional practice.

1. In $\triangle PQR$, $\angle P$ is $65°$, $\angle Q$ is $75°$, and $r = 16$ cm. Find p and q to the nearest centimeter.

2. In $\triangle DEF$, $\angle D$ is $60°$, $\angle E$ is $70°$, and $e = 12$ in. Find d and f to the nearest inch.

3. $n = 12$, $m = 8$, and $\angle N = 50°$. Can MNO be one, two, or no triangle(s)?

4. If MNO in Problem 3 is one or more triangles, find all possible Os to the nearest integer.

Worked Solutions

1. **$p \approx 23$ cm, $q \approx 23$ cm**

    First, find $\angle R$:

    $180° - (65° + 75°) = 180° - 140° = 40°$

    This is an ASA-type problem, so you can use The Law of Sines:

    $$\frac{p}{\sin P} = \frac{r}{\sin R}$$

    Substitute: $\qquad\qquad\qquad \dfrac{p}{\sin 65°} = \dfrac{16}{\sin 40°}$

    Cross multiply: $\qquad\qquad p\sin 40° = 16\sin 65°$

    Look up the sines: $\qquad p(0.643) = 16(0.906)$

    Multiply: $\qquad\qquad\quad p(0.643) = 14.496$

    Divide both sides by 0.643: $\qquad p = 22.54 \approx 23$ cm.

    Now use The Law of Sines to find q: $\quad \dfrac{q}{\sin Q} = \dfrac{r}{\sin R}$

    Substitute: $\qquad\qquad\qquad \dfrac{q}{\sin 75°} = \dfrac{16}{\sin 40°}$

    Cross multiply: $\qquad\qquad q\sin 40° = 16\sin 75°$

    Look up the sines: $\qquad q(0.643) = 16(0.966)$

    Multiply: $\qquad\qquad\quad q(0.643) = 15.456$

    Divide both sides by 0.643: $\qquad q = 24.03 \approx 24$ cm.

2. **$d \approx 11$ in., $f \approx 10$ in.**　　This is an SAA-type problem, so you can use The Law of Sines:

    $$\frac{d}{\sin D} = \frac{e}{\sin E}$$

    Substitute: $\qquad\qquad\qquad \dfrac{d}{\sin 60°} = \dfrac{12}{\sin 70°}$

Cross multiply:	$d\sin 70° = 12\sin 60°$
Look up the sines:	$d(0.94) = 12(0.867)$
Multiply:	$d(0.94) = 10.404$
Divide both sides by 0.94:	$d = 11.07 \approx 11$ in.
Next you need $\angle F$:	$\angle F = 180° - (60° + 70°) = 180° - 130° = 50°$

Now you use The Law of Sines to find f: $\dfrac{d}{\sin D} = \dfrac{f}{\sin F}$

Substitute:	$\dfrac{f}{\sin 50°} = \dfrac{12}{\sin 70°}$
Cross multiply:	$f\sin 70° = 12\sin 50°$
Look up the sines:	$f(0.94) = 12(0.766)$
Multiply:	$f(0.94) = 9.192$
Divide both sides by 0.94:	$f = 9.78 \approx 10$ in.

3. **one**   Since side n is larger than side m, there is only one possible triangle. You may draw the diagram to convince yourself, or you can look back at the preceding four figures.

4. $o \approx 15$ Find o by the Law of Sines, but first you need to find $\angle$s M and O.

Find $\angle M$ by The Law of Sines: $\dfrac{n}{\sin N} = \dfrac{m}{\sin M}$

Substitute:	$\dfrac{12}{\sin 50°} = \dfrac{8}{\sin M}$
Cross multiply:	$12\sin M = 8\sin 50°$
Look up the $\sin 50°$:	$12\sin M = 8(0.766)$
Multiply:	$12\sin M = 6.128$
Divide both sides by 12:	$\sin M = 0.511$
Find the angle whose sine is 0.511:	$M = 30.73 \approx 31°$

$\angle O \approx 180° - (50° + 31°) \approx 99°$. That makes the reference angle for $\angle O$ about 81°.

Now you can find side o by The Law of Sines: $\dfrac{n}{\sin N} = \dfrac{o}{\sin O}$

Substitute:	$\dfrac{12}{\sin 50°} = \dfrac{o}{\sin 81°}$
Cross multiply:	$o\sin 50° = 12\sin 81°$
Look up $\sin 50°$ and $\sin 81°$:	$o(0.766) = 12(0.988)$
Multiply:	$o(0.766) = 11.856$
Divide both sides by 0.766:	$o = 15.48 \approx 15$

Areas of Triangles

The area of any figure is the region within the perimeter of that figure, K, expressed in square units of measure, such as in.^{2} or cm^2. See the following figure. Note that "K" is used for area to avoid confusion with vertex A in $\triangle ABC$.

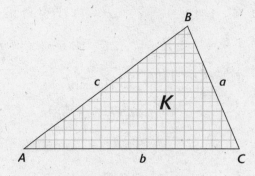

Area is the region inside.

Area for SAS

The most familiar formula for finding the area of a triangle is $K = \frac{1}{2} bh$, but that is by no means the only one. In the preceding figure, if any two sides and the included angle were known (SAS), then area can be found by one of these three formulas:

$$K = \frac{1}{2} bc\sin A, \quad K = \frac{1}{2} ac\sin B, \quad \text{or} \quad K = \frac{1}{2} ab\sin C$$

In other words, area equals half the product of two sides times the sine of the included angle. These formulas are derived from the area formula and a triangle drawn at the origin on a set of axes, but its proof is beyond the scope of this book.

Note: When we use the word "side" in area problems, we're referring to the measure of that side. Also when referring to angles, we're referring to the degree measure of the angles.

Example Problems

These problems show the answers and solutions. Each problem lists certain dimensions of a triangle. Find that triangle's area to the nearest square unit.

1. $a = 16$ cm, $b = 22$ cm, and $\angle C = 30°$

   answer: **88 cm^2**

   $$K = \frac{1}{2} ab\sin C$$

   Substitute: $\quad K = \frac{1}{2} 16 \cdot 22 \sin 30°$

   Find $\sin 30°$: $\quad K = \frac{1}{2} 16 \cdot 22 (0.5)$

Cancel: $K = \frac{1}{2}\cancel{168} \cdot \cancel{22}\,(\cancel{0.5})11$

Multiply: $K = 88$

2. $b = 6$ yd, $c = 8$ yd, $\angle A = 45°$

***answer:* 17 yd.^{2}**

$$K = \frac{1}{2}\,bc\sin A$$

Substitute: $K = \frac{1}{2}\,6 \cdot 8\sin 45°$

Find sin 45°: $K = \frac{1}{2}\,6 \cdot 8\,(0.707)$

Cancel: $K = \frac{\cancel{1}}{\cancel{2}}\cancel{6}3 \cdot 8\,(0.707)$

Multiply: $K = 16.97 \approx 17$

3. $a = 12$ in., $c = 12$ in., and $\angle B = 50°$

***answer:* 55 in.^{2}**

$$K = \frac{1}{2}\,ac\sin B$$

Substitute: $K = \frac{1}{2}\,12 \cdot 12\sin 50°$

Find sin 50°: $K = \frac{1}{2}\,12 \cdot 12\,(0.766)$

Cancel: $K = \frac{\cancel{1}}{\cancel{2}}\cancel{12}6 \cdot 12\,(0.766)$

Multiply: $K = 55.2 \approx 55$

Area for ASA or SAA

Using the Law of Sines and substituting into the SAS area formulas leads to the development of area formulas for situations in which two angles and a side, included or not, are known. Those formulas are

$$K = \frac{1}{2}\,a^2\,\frac{\sin B\sin C}{\sin A}$$

$$K = \frac{1}{2}\,b^2\,\frac{\sin A\sin C}{\sin B}$$

$$K = \frac{1}{2}\,c^2\,\frac{\sin A\sin B}{\sin C}$$

All three formulas really state the same thing, depending upon which side you know. Area = half the square of the known side times the sines of the angles opposite the two other sides divided by the sine of the angle opposite the known side.

Example Problems

These problems show the answers and solutions. Each problem lists certain dimensions of a triangle. Find that triangle's area to the nearest square unit.

1. $\angle A = 120°$, $\angle C = 30°$, $a = 7$ in.

   answer: 7 in.2

   $\angle B = 180° - (120° + 30°) = 180° - 150° = 30°$

   This is an SAA problem using a: $K = \dfrac{1}{2} a^2 \dfrac{\sin B \sin C}{\sin A}$

   Substitute: $K = \dfrac{1}{2} 7^2 \dfrac{\sin 30° \sin 30°}{\sin 120°}$

   Evaluate: $K = \dfrac{1}{2}(49) \dfrac{(.5)(.5)}{0.866}$

   Multiply: $K = (24.5) \dfrac{(.5)(.5)}{0.866} = \dfrac{6.125}{0.866}$

   Finally, divide by 0.866: $K = \dfrac{6.125}{0.866} = 7.072 \approx 7$

2. $\angle A = 55°$, $\angle B = 65°$, $c = 10$ ft.

   answer: 43ft.²

   $\angle C = 180° - (65° + 55°) = 180° - 120° = 60°$

   This is an ASA problem using b: $K = \dfrac{1}{2} c^2 \dfrac{\sin A \sin B}{\sin C}$

   Substitute: $K = \dfrac{1}{2} 10^2 \dfrac{\sin 55° \sin 65°}{\sin 60°}$

   Evaluate: $K = \dfrac{1}{2}(100) \dfrac{(0.819)(0.906)}{0.866}$

   Multiply: $K = (50) \dfrac{(0.819)(0.906)}{0.866} = \dfrac{37.1007}{0.866}$

   Finally, divide by 0.866: $K = \dfrac{37.1007}{0.866} = 42.84 \approx 43$

3. $\angle A = 40°$, $\angle B = 50°$, $b = 12$ cm

   answer: 60 cm²

   This looks like an SAA problem, but there's more. This is a right triangle with the right angle at C. If you use b for the base, (AC) then BC is the height. You can find h by using $\cos\angle B$. Look at the following figure.

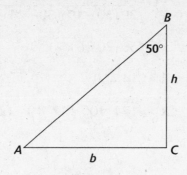

Use the tan ratio:	$\tan B = \dfrac{b}{h}$
Substitute:	$\tan 50° = \dfrac{12}{h}$
Cross multiply:	$h \tan 50° = 12$
Find the value of $\tan 50°$:	$h(1.192) = 12$
Divide both sides by 1.192:	$h = 10.067$. Call it 10.
Now use the standard area formula:	$K = \dfrac{1}{2} bh$
Substitute:	$K = \dfrac{1}{2} 12 \cdot 10$
Solve:	$K = 60$

Heron's Formula (SSS)

The ancient Greek philosopher and mathematician Heron (also known as Hero) developed a formula for figuring out the area of triangles given only the lengths of the three sides. If a, b, and c are the three sides of a triangle, and s is the semiperimeter (meaning half the perimeter) $\dfrac{a+b+c}{2}$, then

$$K = \sqrt{s(s-a)(s-b)(s-c)}$$

There are many different proofs of Heron's Formula, but again, that is beyond the scope of this volume. There is a very interesting one in *CliffsQuickReview: Trigonometry*.

Example Problems

These problems show the answers and solutions. Use Heron's Formula to find the areas of the three triangles. The answers may be left in radical form.

1. $a = 10$ cm, $b = 24$ cm, $c = 26$ cm.

   answer: **120 cm²**

   First, state Heron's Formula: $K = \sqrt{s(s-a)(s-b)(s-c)}$

   $$s = \frac{1}{2}(10 + 24 + 26) = 30$$

Substitute:　　　　$K = \sqrt{30(30-10)(30-24)(30-26)}$

Subtract:　　　　　$K = \sqrt{30(20)(6)(4)}$

Multiply:　　　　　$K = \sqrt{14,400}$

Simplify:　　　　　$K = \sqrt{144 \cdot 100} = 12 \cdot 10 = 120$

2.　$a = 9$ in., $b = 12$ in., $c = 15$ in.

answer: 54 in.^{2}

First, state Heron's Formula:　$K = \sqrt{s(s-a)(s-b)(s-c)}$

$$s = \frac{1}{2}(9+12+15) = 18$$

Substitute:　　　　$K = \sqrt{18(18-9)(18-12)(18-15)}$

Subtract:　　　　　$K = \sqrt{18(9)(6)(3)}$

Multiply:　　　　　$K = \sqrt{2916}$

Simplify:　　　　　$K = 54$

3.　$a = 30$ m, $b = 16$ m, $c = 34$ m.

answer: 240 m^2

First, state Heron's Formula:　$K = \sqrt{s(s-a)(s-b)(s-c)}$

$$s = \frac{1}{2}(30+16+34) = 40$$

Substitute:　　　　$K = \sqrt{40(40-30)(40-16)(40-34)}$

Subtract:　　　　　$K = \sqrt{40(10)(24)(6)}$

Multiply:　　　　　$K = \sqrt{57,600}$

Simplify:　　　　　$K = 240$

Work Problems

Use these problems to give yourself additional practice. Each problem lists certain dimensions of a triangle. First, recognize what formula must be used to find that triangle's area and then find the triangle's area.

1.　$b = 12$ m, $c = 18$ m, and $\angle A = 40°$

2.　$\angle A = 75°$, $\angle B = 35°$, and $c = 12$ cm

3.　$a = 30$ cm, $b = 15$ cm, and $c = 25$ cm

4.　$\angle A = 60°$, $\angle C = 70°$, and $b = 9$

Worked Solutions

1. **69 m²**   This is an SAS problem, so $K = \frac{1}{2} bc\sin A$

    Substitute: $K = \frac{1}{2} 12 \cdot 18 \sin 40°$

    Find sin 40°: $K = \frac{1}{2} 12 \cdot 18 (0.643)$

    Cancel: $K = \frac{1}{2} \cancel{12}6 \cdot 18 (0.643)$

    Multiply: $K = 69.444 \approx 69$

2. **42 cm²** $\angle C = 180° - (75° + 35°) = 180° - 110° = 70°$

    This is an ASA problem using b: $K = \frac{1}{2} c^2 \dfrac{\sin A \sin B}{\sin C}$

    Substitute: $K = \frac{1}{2} 12^2 \dfrac{\sin 75° \sin 35°}{\sin 70°}$

    Evaluate: $K = \frac{1}{2}(144) \dfrac{(0.966)(0.574)}{0.94}$

    Multiply: $K = (72)\dfrac{(0.966)(0.574)}{0.94} = \dfrac{39.92}{094}$

    Finally, divide by 0.866: $K = \dfrac{39.92}{0.94} = 42.47 \approx 42$

3. **$50\sqrt{14}$ cm²**   This is an SSS situation, so use Heron's Formula:

    $$K = \sqrt{s(s-a)(s-b)(s-c)}$$

    Find the semiperimeter: $s = \frac{1}{2}(30 + 15 + 25) = 35$

    Substitute: $K = \sqrt{35(35-30)(35-15)(35-25)}$

    Subtract: $K = \sqrt{35(5)(20)(10)}$

    Multiply: $K = \sqrt{35,000}$

    Simplify: $K = \sqrt{2500 \cdot 14} = 50\sqrt{14}$

4. **43 units²**   This is an SAA problem using $K = \frac{1}{2} b^2 \dfrac{\sin A \sin C}{\sin B}$

    $\angle B = 180° - (60° + 70° = 180° - 130° = 50°)$

    Substitute: $K = \frac{1}{2} 9^2 \dfrac{\sin 60° \sin 70°}{\sin 50°}$

    Evaluate: $K = \frac{1}{2}(81) \dfrac{(0.866)(0.94)}{0.766}$

    Multiply: $K = (40.5) \dfrac{(0.866)(0.94)}{0.766} = \dfrac{32.969}{0.766}$

    Finally, divide by 0.766: $K = \dfrac{32.969}{0.766} = 43.04 \approx 43$

Chapter Problems and Solutions

Problems

Solve these problems for more practice applying the skills from this chapter. Worked out solutions follow problems.

1. In right triangle $\triangle PQR$, $m\angle P = 50°$ and hypotenuse $q = 15$. Solve the triangle.

2. In right triangle $\triangle ABC$, $m\angle A = 38°$ and C is the right angle. $b = 20$. Solve the triangle.

3. Kira sat on top of a mountain on a cloudless day. She looked out toward the road below at an angle of depression of 65° through an optical rangefinder. The rangefinder showed the distance to the road as being 6,000 feet. How high above the road was Kira sitting? The rangefinder she was looking through was 3 feet above the ground.

4. Ian's kite is on the end of a 100-foot string flying at an angle of elevation of 55°. The hand in which Ian is holding the string is 4 feet above the ground. How high is the kite?

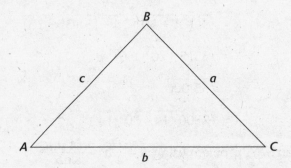

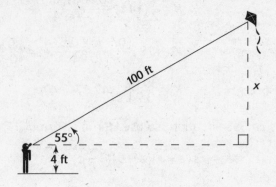

5. Solve the triangle in the preceding figure if $m\angle A = 45°$, $m\angle C = 65°$, and $b = 12$.

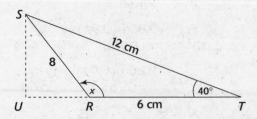

6. Using the Law of Sines and triangle *RST* in the preceding figure, find *m∠SRT*.

7. In △*ABC*, *a* = 8, *b* = 9, and *c* = 6. To the nearest degree, find ∠*C*.

8. In △*ABC*, *a* = 14, *b* = 10, and *c* = 6. Find ∠*A*.

9. In △*RST*, *r* = 13, *s* = 10, and *t* = 11. Find the angles to the nearest whole degree.

10. In △*EFG*, *e* = 5, *f* = 11, and *g* = 12. Find the angles to the nearest whole degree.

11. In △*DEF*, *e* = 8, *f* = 10, and ∠*D* = 60°. Solve the triangle, rounding all lengths and angles to the nearest whole unit.

12. In △*ABC*, *a* = 12, *b* = 12, and ∠*C* = 50°. Solve the triangle, rounding all lengths and angles to the nearest whole units.

13. A radio tower stands atop a hill that is sloped 20° from the horizontal. An observer, lying on the ground 120 feet down the hill sights the top of that tower at an angle of elevation of 25°. Assuming that observation to be at eye level, how tall to the nearest foot is the tower?

14. In △*PQR*, ∠*P* is 35°, ∠*Q* is 65°, *r* = 16 cm. Find *p* and *q* to the nearest centimeter.

15. In △*DEF*, ∠*D* is 60°, ∠*E* is 70°, *e* = 12 in. Find *d* and *f* to the nearest inch.

16. Two observers on level ground spot a balloon in the air at exactly the same moment. One observer sees the angle of elevation as 67°, and the other observer, who is 5,280 feet away from the first, sees the elevation as 70°. Find the height of the balloon.

17. Find the number of triangles that can be formed if ∠*D* = 115°, *e* = 16, and *d* = 15.

18. Find the number of triangles that can be formed if ∠*D* = 90°, *f* = 30, and *d* = 35.

19. In △*ABC*, *b* = 8 m, *c* = 9 m, and ∠*A* = 45°. Find the triangle's area to the nearest square unit.

20. In △*ABC*, *a* = 10 cm, *c* = 14 cm, and ∠*B* = 50°. Find the triangle's area to the nearest square unit.

Each problem (21–25) lists certain dimensions of a triangle. Find that triangle's area to the nearest square unit.

21. ∠*A* = 110°, ∠*C* = 35°, and *a* = 8 in.

22. ∠*D* = 64°, ∠*E* = 72°, and *f* = 20 ft.

23. ∠*G* = 67°, ∠*I* = 74°, and *h* = 12 cm.

24. *a* = 12 cm, *b* = 14 cm, and *c* = 16 cm.

25. *a* = 18 in., *b* = 24 in., and *c* = 30 in.

Answers and Solutions

1. **Answer:** $m\angle Q = 90°$, $m\angle R = 40°$, $p = 11.5$, $r = 9.65$.　You know that the right angle is $\angle Q$ since q is the hypotenuse. $90° + 50° = 140°$, so $m\angle R = 180° - 140° = 40°$. You can find the sides by means of sin, and/or cosine, and/or secant, and/or cosecant. The easiest is probably as follows:

$$\sin R = \frac{r}{q} \qquad\qquad \sin P = \frac{p}{q}$$

$$\begin{aligned} r &= q\sin R & p &= q\sin P \\ r &= 15\sin 40° & p &= 15\sin 50° \\ r &= 15(0.643) & p &= 15(0.766) \\ r &= 9.65 & p &= 11.5 \end{aligned}$$

2. **Answer:** $m\angle B = 52°$, $a = 4.66$, $c = 11.04$.　$90° + 38° = 128°$, so $m\angle B = 180° - 128° = 52°$. There are many ways to solve these, and the following is just one:

$$\tan A = \frac{a}{b} \qquad\qquad \sin B = \frac{b}{c}$$

$$\begin{aligned} a &= b\tan A & b &= c\sin B \\ a &= 20\tan 38° & c &= \frac{b}{\sin B} \\ a &= 20(0.781) & c &= \frac{20}{\sin 52°} \\ a &= 15.6 & c &= \frac{20}{0.788} \\ & & c &= 25.4 \end{aligned}$$

3. **Answer:** 5,433 ft.　Draw your own diagram. Note Kira's line of sight is the hypotenuse. Her height above the road is the side oposite the angle of depression.

$$\sin 65° = \frac{x}{6000}$$

$$\begin{aligned} x &= (6000)(\sin 65°) \\ x &= (6000)(0.906) \\ x &= 5436 \text{ ft.} \end{aligned}$$

But she is holding the range finder 3 feet above the ground, so subtract that from the total to get 5,433 ft.

4. **Answer:** 85.9 ft.　First, picture the problem as in the following figure.

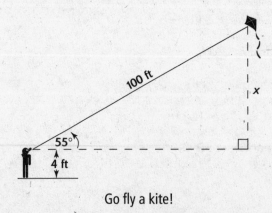

Go fly a kite!

Here, you know the adjacent side of the triangle (with respect to the angle of elevation) and want to find the opposite side, so use tangent:

$$\sin 55° = \frac{x}{100}$$
$$x = 100(\sin 55°)$$
$$x = 100(0.819)$$
$$x = 81.9 \text{ ft.}$$

Add 4 feet for the height of Ian's hand and get 85.9 ft.

5. ***Answer:*** $m\angle B = 70°$, $a = 13.04$, $c = 14.74$.   Since the total sum of the angles = 180°, and $45° + 65° = 110°$, $m\angle B = 180 - 110 = 70°$.

From the Law of Sines, you get $\frac{a}{\sin A} = \frac{b}{\sin B} = \frac{c}{\sin C}$

That means $\frac{a}{\sin 45°} = \frac{12}{\sin 70°} = \frac{c}{\sin 65°}$

To solve, make two proportions:

$\frac{a}{\sin 45°} = \frac{12}{\sin 70°}$ and $\frac{12}{\sin 70°} = \frac{c}{\sin 65°}$

Solve each.

$$\frac{12}{0.940} = \frac{c}{0.906}$$

$0.940a = 12(0.707)$ $0.940c = 12(0.906)$

$0.940a = 8.484$ $0.940c = 10.872$

$a = 9.03$ $c = 11.57$

6. ***Answer:*** 105°.   First, set up the equation, $\frac{r}{\sin x} = \frac{t}{\sin T}$.

Use x ($\angle SRT$) because it's the angle for which you are looking and $\angle T$ because it's the only angle whose measure you know. Of course, you know all three sides.

$$\frac{12}{\sin x} = \frac{8}{\sin 40}$$

Cross multiply: $8(\sin x) = 12 \sin 40°$

Find $\sin 40°$: $8(\sin x) = 12(0.643)$

Multiply: $8(\sin x) = 7.716$

Divide both sides by 8: $\sin x = 0.965$

Find arcsin 0.965: $\angle SRT = 74.8° \approx 75°$

But $\angle SRT$ is an obtuse angle! Arcsin and sin, for that matter, only exist for angles up to 90°. So what did you actually find? You found the measure of $\angle SRU$, the quadrant II reference angle for $\angle SRT$. To get $m\angle SRT$, subtract $\angle SRU$ from 180°. $180° - 75° = 105°$, which is the answer.

7. **Answer:** 41°

Use the $\cos C =$ form: $\quad \cos C = \dfrac{a^2 + b^2 - c^2}{2ab}$

Substitute: $\quad \cos C = \dfrac{8^2 + 9^2 - 6^2}{2 \cdot 8 \cdot 9}$

Multiply: $\quad \cos C = \dfrac{64 + 81 - 36}{2 \cdot 72}$

Combine like terms: $\quad \cos C = \dfrac{109}{144} = 0.757$

Find arccos 0.757: $\quad \angle C = 40.8° \approx 41°$

8. **Answer:** 120°

Use the $\cos A =$ form: $\quad \cos A = \dfrac{b^2 + c^2 - a^2}{2bc}$

Substitute: $\quad \cos A = \dfrac{10^2 + 6^2 - 14^2}{2 \cdot 10 \cdot 6}$

Multiply: $\quad \cos A = \dfrac{100 + 36 - 196}{2 \cdot 60}$

Combine like terms: $\quad \cos A = \dfrac{-60}{120} = -\dfrac{1}{2} = -.5$

Find arccos: −0.500

You need to look up arccos 0.500, which finds $\angle C = 60°$, but you're not done. That negative value for cosine tells you that this angle is in quadrant II, so it's an obtuse angle obtained by subtracting the angle you found from 180°: $180° - 60° = 120°$.

9. **Answer:** $m\angle R \approx 76°$, $m\angle S \approx 48°$, $m\angle T \approx 56°$

The largest angle is $\angle R$: $\quad \cos R = \dfrac{s^2 + t^2 - r^2}{2st}$

Substitute: $\quad \cos R = \dfrac{10^2 + 11^2 - 13^2}{2(10)(11)}$

Multiply: $\quad \cos R = \dfrac{100 + 121 - 169}{2(110)}$

Combine like terms: $\quad \cos R = \dfrac{52}{220} = 0.236$

Find arccos 0.236: $\quad R = 76.34 \approx 76°$

On to the Law of Sines: $\quad \dfrac{r}{\sin R} = \dfrac{s}{\sin S}$

Substitute: $\quad \dfrac{13}{\sin 76°} = \dfrac{10}{\sin S}$

Cross multiply: $\quad 13 \sin S = 10(0.97)$

Multiply: $\quad 13 \sin S = 9.7$

Divide both sides by 13: $\quad \sin S = 0.746$

Find arcsin 0.746:   48.24°, so $m\angle S \approx 48°$

$m\angle T \approx 180° - (76° + 48°) \approx 56°$

10. ***Answer:*** $m\angle G \approx 89°$, $m\angle E \approx 25°$, $m\angle F \approx 66°$.   The largest angle is $\angle G$:

$$\cos G = \frac{e^2 + f^2 - g^2}{2ef}$$

Substitute: $\cos G = \frac{5^2 + 11^2 - 12^2}{2 \cdot 5 \cdot 11}$

Multiply: $\cos G = \frac{25 + 121 - 144}{2 \cdot 55}$

Combine like terms: $\cos G = \frac{2}{110} = \frac{1}{55} = 0.018$

Find arccos 0.018: $G = 88.9° \approx 89°$

On to the Law of Sines: $\frac{g}{\sin G} = \frac{e}{\sin E}$

Substitute: $\frac{12}{\sin 89} = \frac{5}{\sin E}$

Look up sin89°: $\frac{12}{0.999} = \frac{5}{\sin E}$

Cross multiply: $12\sin E = 5(.0999)$

Multiply: $12\sin E = 4.995$

Divide both sides by 12: $\sin E = 0.416$

Find arcsin 0.416:   24.58°, so $m\angle E \approx 25°$

$m\angle F \approx 180° - (89° + 25°) \approx 66°$

11. ***Answer:*** $d \approx 9$, $m\angle E \approx 50°$, $m\angle F = 70°$.   You need to find d first.

$$\cos D = \frac{e^2 + f^2 - d^2}{2ef}$$

Substitute: $\frac{0.5}{1} = \frac{8^2 + 10^2 - d^2}{2(8)(10)}$

Multiply: $\frac{0.5}{1} = \frac{64 + 100 - d^2}{2(80)}$

Combine like terms: $\frac{0.5}{1} = \frac{164 - d^2}{160}$

Cross multiply: $164 - d^2 = 0.5 \cdot 160$

Combine like terms: $-d^2 = 0.5 \cdot 160 - 164 = 80 - 164 = -84$

Take the square root of both sides: $d = \sqrt{84} = 2\sqrt{21} \approx 9.165 \approx 9$

Now use the Law of Sines to find e: $\dfrac{e}{\sin E} = \dfrac{d}{\sin D}$

Substitute: $\dfrac{8}{\sin E} = \dfrac{9}{\sin 60°}$

Find sin 60°: $\dfrac{8}{\sin E} = \dfrac{9}{0.866}$

Cross multiply: $9\sin E = 8(0.866)$

Multiply: $9\sin E = 6.928$

Divide both sides by 9: $\sin E = 0.77$

Find arcsin 0.77: $E = 50.35°$, so $m\angle E \approx 50°$

$m\angle F \approx 180° - (60° + 50°) \approx 70°$

12. **Answer:** $m\angle A = 65°$ $m\angle B = 65°$, $c \approx 10$

Since legs a and b are equal in length, $\triangle ABC$ is an isosceles triangle, so both angles A and B are equal to half the difference of $180° - 50°$. $180° - 50° = 130°$, and half that is 65°.

To find side c, use the Law of Sines: $\dfrac{c}{\sin C} = \dfrac{a}{\sin A}$

Substitute: $\dfrac{c}{\sin 50°} = \dfrac{12}{\sin 65°}$

Cross multiply: $c\sin 65° = 12\sin 50°$

Look up the sines: $c(0.906) = 12(0.766)$

Multiply: $c(0.906) = 9.192$

Divide both sides by .906: $c = 10.146 \approx 10$

13. **Answer:** 72 ft.　Drawing the diagram shown in the following figure will help you see that this is an ASA problem, with the angle formed between the base of the tower and the observer at $110°$ $(90° + 20° = 110°)$. That makes $\angle Z = 180° - (110° + 25°) = 180° - 135° = 45°$.

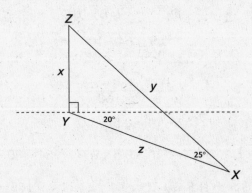

Find the tower's height by The Law of Sines: $\dfrac{x}{\sin X} = \dfrac{z}{\sin Z}$

Substitute: $\dfrac{x}{\sin 25°} = \dfrac{120}{\sin 45°}$

Cross multiply: $x\sin 45° = 120\sin 25°$

Look up the sines: $x(0.707) = 120(0.423)$

Multiply: $x(0.707) = 50.76$

Divide both sides by 0.707: $x = 71.79 \approx 72$ ft.

14. **Answer:** $p \approx 9$ cm, $q \approx 15$ cm. First, find $\angle R$.

$180° - (35° + 65°) = 180° - 100° = 80°$

This is an ASA-type problem, so you can use The Law of Sines: $\dfrac{p}{\sin P} = \dfrac{r}{\sin R}$

Substitute: $\dfrac{p}{\sin 35°} = \dfrac{16}{\sin 80°}$

Cross multiply: $p\sin 80° = 16\sin 35°$

Look up the sines: $p(0.985) = 16(0.574)$

Multiply: $p(0.985) = 9.184$

Divide both sides by 0.985: $p = 9.32 \approx 9$ cm

Now you use The Law of Sines to find q: $\dfrac{q}{\sin Q} = \dfrac{r}{\sin R}$

Substitute: $\dfrac{q}{\sin 65°} = \dfrac{16}{\sin 80°}$

Cross multiply: $q\sin 80° = 16\sin 65°$

Look up the sines: $q(0.985) = 16(0.906)$.

Multiply: $q(0.985) = 14.496$.

Divide both sides by 0.985: $q = 14.71 \approx 15$ cm.

15. **Answer:** $d \approx 11$ in., $f \approx 10$ in. This is an SAA-type problem, so you can use The Law of Sines:

$$\dfrac{d}{\sin D} = \dfrac{e}{\sin E}$$

Substitute: $\dfrac{d}{\sin 60°} = \dfrac{12}{\sin 70°}$

Cross multiply: $d\sin 70° = 12\sin 60°$

Look up the sines: $d(0.94) = 12(0.867)$

Multiply: $d(0.94) = 10.404$

Divide both sides by 0.94: $d = 11.07 \approx 11$ in.

Next you need. $\angle F = 180° - (60° + 70°) = 180° - 130° = 50°$

Now use The Law of Sines to find f: $\dfrac{e}{\sin E} = \dfrac{f}{\sin F}$

Substitute: $\dfrac{f}{\sin 50°} = \dfrac{12}{\sin 70°}$

Cross multiply: $f \sin 70° = 12 \sin 50°$

Look up the sines: $f(0.94) = 12(0.766)$

Multiply: $f(0.94) = 9.192$

Divide both sides by 0.94: $f = 9.78 \approx 10$ in.

16. **Answer:** 6,876 ft.

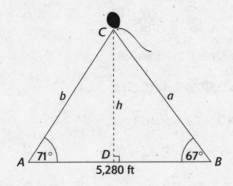

Find the third angle: $180° - (71° + 67°) = 42°$

Use the Law of Sines to find side a: $\dfrac{a}{\sin A} = \dfrac{c}{\sin C}$

Substitute: $\dfrac{a}{\sin 71°} = \dfrac{5280}{\sin 42°}$

Cross multiply: $a(\sin 42°) = 5280(\sin 71°)$

Collect terms: $a = \dfrac{5280(\sin 71°)}{\sin 42°}$

Look up sines: $a = \dfrac{5280(0.946)}{0.669}$

Compute: $a = 7466$ ft.

It's no accident that a is the hypotenuse of right $\triangle BCD$. To find the height of the balloon, use $\sin 67°$.

Set up: $\sin 67° = \dfrac{h}{a}$

Substitute: $\sin 67° = \dfrac{h}{7466}$

Collect terms: $h = 7466 \sin 67°$

Look up $\sin 67°$: $h = 7466 \cdot 0.921$

Compute: $h = 6876$ ft.

17. **Answer:** None

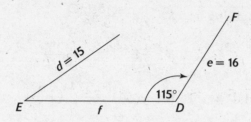

Notice the representation in the preceding figure. Since *d* is shorter than *e*, there can be no triangle.

18. **Answer:** 1.   Since $\angle D = 90°$, and the side opposite it is longer than the side adjacent to it, exactly one triangle can be formed. See the following figure.

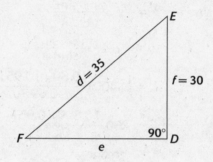

19. **Answer:** 25 m²

Use the SAS form with *b* and *c*: $K = \frac{1}{2} bc \sin A$

Substitute: $K = \frac{1}{2} 8 \cdot 9 \sin 45°$

Find $\sin 45°$: $K = \frac{1}{2} 8 \cdot 9 (0.707)$

Cancel: $K = \frac{1}{2} 8\,4 \cdot 9 (0.707)$

Multiply: $K = 25.452 \approx 25$

20. **Answer:** 54 cm²

Use SAS form with *a* and *c*: $K = \frac{1}{2} ac \sin B$

Substitute: $K = \frac{1}{2} 10 \cdot 14 \sin 50°$

Find $\sin 50°$: $K = \frac{1}{2} 10 \cdot 14 (0.766)$

Cancel: $K = \frac{1}{2} 10\,5 \cdot 14 (0.766)$

Multiply: $K = 53.62 \approx 54$

21. ***Answer:*** 11 in.^2　　$\angle B = 180° - (110° + 35°) = 180° - 145° = 35°$

    This is an SAA problem using a:　$K = \frac{1}{2} a^2 \frac{\sin B \sin C}{\sin A}$

    Substitute:　　　　　　　　$K = \frac{1}{2} 8^2 \frac{\sin 35° \sin 35°}{\sin 110°}$

    Evaluate (Use 70° for 110°):　$K = \frac{1}{2}(64)\frac{(0.574)(0.574)}{0.94}$

    Multiply:　　　　　$K = (32)\frac{(0.574)(0.574)}{0.94} = \frac{10.543}{0.94}$

    Finally, divide by 0.0.94:　$K = \frac{10.543}{0.94} = 11.22 \approx 11$

22. ***Answer:*** 246 ft.^2　$\angle F = 180° - (64° + 72°) = 180° - 136° = 44°$

    This is an ASA problem using f:　$K = \frac{1}{2} f^2 \frac{\sin D \sin E}{\sin F}$

    Substitute:　　　　　　　　$K = \frac{1}{2} 20^2 \frac{\sin 64° \sin 72°}{\sin 44°}$

    Evaluate:　　　　　$K = \frac{1}{2}(400)\frac{(0.899)(0.951)}{0.695}$

    Multiply:　　　　$K = (200)\frac{(0.899)(0.951)}{0.695} = \frac{170.9898}{0.695}$

    Finally, divide by 0.695:　$K = \frac{170.9898}{0.695} = 246.03 \approx 246$

23. ***Answer:*** 101 cm^2　$\angle H = 180° - (67° + 74°) = 180° - 141° = 39°$

    This is an SAA problem using h:　$K = \frac{1}{2} h^2 \frac{\sin G \sin I}{\sin H}$

    Substitute:　　　　　　　　$K = \frac{1}{2} 12^2 \frac{\sin 67° \sin 74°}{\sin 39°}$

    Evaluate:　　　　　$K = \frac{1}{2}(144)\frac{(0.921)(0.961)}{0.629}$

    Multiply:　　　　$K = (72)\frac{(0.921)(0.961)}{0.629} = \frac{63.726}{0.629}$

    Finally, divide by 0.629:　$K = \frac{63.726}{0.629} = 101.313 \approx 101$

24. ***Answer:*** $21\sqrt{15} \text{ cm}^2$

    It's SSS, so use Heron's Formula:　$K = \sqrt{s(s-a)(s-b)(s-c)}$

    Find the semiperimeter:　$s = \frac{1}{2}(12 + 14 + 16) = 21$

    Substitute:　$K = \sqrt{21(21-12)(21-14)(21-16)}$

    Subtract:　$K = \sqrt{21(9)(7)(5)}$

    Multiply:　$K = \sqrt{6615}$

    Simplify:　$K = \sqrt{6615} = \sqrt{9 \cdot 49 \cdot 15} = 21\sqrt{15}$

25. **Answer:** 216 in.^{2}

   It's SSS, so use Heron's Formula: $K = \sqrt{s(s-a)(s-b)(s-c)}$

   Find the semiperimeter: $s = \frac{1}{2}(18 + 24 + 30) = 36$

   Substitute: $K = \sqrt{36(36-18)(36-24)(36-30)}$

   Subtract: $K = \sqrt{36(18)(12)(6)}$

   Multiply: $K = \sqrt{46,656}$

   Simplify: $K = 216$

Supplemental Chapter Problems

Solve these problems for even more practice applying the skills from this chapter. The Answer section will direct you to where you need to review.

Problems

1. In right triangle $\triangle PQR$, $m\angle P = 30°$ and hypotenuse $q = 12$. Solve the triangle.

2. In right triangle $\triangle ABC$, $m\angle A = 36°$ and C is the right angle. $b = 16$. Solve the triangle.

3. The top of a ladder is leaning against a building. Its bottom is 5 feet away from the base of the building. The ladder forms a 78° angle of elevation with the ground. To the nearest foot, how long is the ladder?

4. Willis stood on top of a cliff on a cloudless day. He looked out toward the lake below at an angle of depression of 60° through an optical rangefinder held 5 feet above the ground on which he was standing. The rangefinder showed the distance to the valley floor to be 7,500 feet away. How high above the lake's surface was Willis standing?

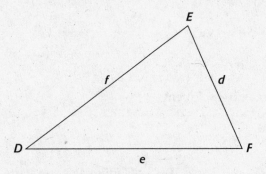

Use the preceding figure as the model for Problems 5 and 6.

5. If $m\angle D = 75°$, $m\angle E = 55°$, and $d = 18$ cm, find the length of f to the nearest integer.

6. If $m\angle E = 80°$, $e = 16$ cm, and $f = 12$ cm, find $m\angle F$ to the nearest degree.

7. In $\triangle ABC$, $b = 10$, $c = 8$, and $\cos A = \frac{3}{8}$. Find a to the nearest integer.

8. In △ABC, a = 12, b = 14, and c = 16. Find ∠B to the nearest degree.

9. In △RST, r = 23, s = 21, and t = 32. Find the angles to the nearest whole degree.

10. In △MNO, m = 18, n = 12, and o = 14. Find the angles to the nearest whole degree.

11. In △DEF, e = 10, f = 12, and ∠D = 75°. Solve the triangle rounding all lengths and angles to the nearest half unit.

12. In △PQR, p = 7, q = 6, and ∠R = 85°. Solve the triangle rounding all lengths and angles to the nearest half unit.

13. In △HIJ, ∠H is 50°, ∠J is 70°, and i = 10 in. Find h and j to the nearest inch.

14. In △PQR, ∠P is 55°, ∠Q is 65°, and r = 20 cm. Find p and q to the nearest centimeter.

15. In △DEF, m∠D is 110°, m∠E is 40°, and e = 14 in. Find d and f to the nearest inch.

16. In △ABC, ∠A is 74°, ∠B is 63°, and b = 24 cm. Find a and c to the nearest cm.

17. Find the number of triangles that can be formed if ∠A = 120°, b = 16, and a = 12.

18. Find the number of triangles that can be formed if ∠A = 90°, b = 25, and a = 40.

19. In △MNO, m = 10 yd., n = 12 yd., ∠O = 60°. Find the triangle's area to the nearest square unit.

20. In △RST, s = 20 in., t = 24 in., and ∠R = 37°. Find the triangle's area to the nearest square unit.

Each problem (21–25) lists certain dimensions of a triangle. Find that triangle's area to the nearest square unit.

21. ∠A = 80°, ∠C = 40°, and a = 10 in.

22. ∠D = 69°, ∠E = 65°, and f = 16 ft.

23. ∠G = 44°, ∠I = 64°, and h = 12 cm.

24. a = 12 in., b = 13 in., and c = 15 in.

25. a = 18 cm, b = 16 cm, and c = 24 cm

Answers

1. $m\angle Q = 90°$, $m\angle R = 60°$, $p = 6$, $r = 10.4$ (Finding Missing Parts of Right Triangles, p. 93)

2. $m\angle B = 54°$, $a = 11.6$, $c = 19.8$ (Finding Missing Parts of Right Triangles, p. 93)

3. 24 ft.   (Angles of Elevation and Depression, p. 95)

4. 6,490 ft.   (Angles of Elevation and Depression, p. 95)

5.   14   (The Law of Sines, p. 100)

6.   48°   (The Law of Sines, p. 100)

7.   10   (The Law of Cosines, p. 105)

8.   58°   (The Law of Cosines, p. 105)

9. $R \approx 52°$, $S \approx 41°$, $T \approx 87°$ (Solving General Triangles—SSS, p. 109)

10. $M \approx 87°$, $N \approx 42°$, $O \approx 51°$ (Solving General Triangles—SSS, p. 109)

11. $d \approx 13.5$, $m\angle E \approx 45.5°$, $m\angle F \approx 59.5°$ (Solving General Triangles—SAS, p. 109)

12. $r \approx 9$, $m\angle P \approx 51°$, $m\angle Q \approx 44°$ (Solving General Triangles—SAS, p. 109)

13. $h \approx 9$, $j \approx 11$ (Solving General Triangles—ASA, p. 111)

14. $p \approx 19$, $q \approx 21$ (Solving General Triangles—ASA, p. 111)

15. $d \approx 20$, $f \approx 11$ (Solving General Triangles—SAA, p. 111)

16. $a \approx 26$, $c \approx 18$ (Solving General Triangles—SAA, p. 111)

17.   None   (Solving General Triangles—SSA, The Ambiguous Case, p. 111)

18.   One   (Solving General Triangles—SSA, The Ambiguous Case, p. 111)

19.   52 yd.^{2} (Area for SAS, p. 116)

20.   144 in.^{2} (Area for SAS, p. 116)

21.   28 in.^{2} (Area for ASA or SAA, p. 117)

22.   151 ft.^{2} (Area for ASA or SAA, p. 117)

23.   47 cm^2 (Area for ASA or SAA, p. 117)

24.   75 in.^{2} (Heron's Formula [SSS], p. 119)

25.   144 cm^2 (Heron's Formula [SSS], p. 119)

Chapter 4
Trigonometric Identities

Some equations are valid for only a single value of the variable. Such equations are known as **conditional equations,** and they may be true for one replacement of the variable, no replacements of the variable, or many replacements of the variable, but they are never true for *all* replacements of the variable. For example, $2x = 6$ is a conditional equation. It is only true for $x = 3$.

Fundamental Identities

There are also equations that are valid for all values of the variable within the defined solution set. That type of equation is known as an **identity.** For example, $5x = 3x + 2x$ is always true. $\tan\theta = \dfrac{\sin\theta}{\cos\theta}$ is an identity for all real values except $\cos\theta = 0$.

Identities are useful for solving more complex trigonometric equations by distilling them down into more recognizable forms. If an equation can be shown to be false for any single value of its variable, then that is enough to prove that it is not an identity. Such a value is known as a **counterexample.**

The fundamental identities can be separated into several groups.

Reciprocal Identities

You are already familiar with the **reciprocal identities.** They are

$$\frac{1}{\sin\theta} = \csc\theta$$

$$\frac{1}{\cos\theta} = \sec\theta$$

$$\frac{1}{\tan\theta} = \cot\theta$$

$$\frac{1}{\cot\theta} = \tan\theta$$

$$\frac{1}{\sec\theta} = \cos\theta$$

$$\frac{1}{\csc\theta} = \sin\theta$$

The most important thing you need to remember about the reciprocal identities is which is the reciprocal of which. Remembering that secant and cosine go together is usually enough to be able to sort the list correctly.

Ratio Identities

The **ratio identities** are also sometimes referred to as the **quotient identities.** They are

$$\tan\theta = \frac{\sin\theta}{\cos\theta}$$

$$\cot\theta = \frac{\cos\theta}{\sin\theta}$$

Until you have used them to simplify complex trigonometric equations, you cannot appreciate how useful these two identities can be.

Example Problems

These problems show the solutions.

1. Prove that $\tan\theta = \frac{\sin\theta}{\cos\theta}$.

    answer: By definition, in right $\triangle ABC$, with the right angle at C and θ at A, $\sin\theta = \frac{a}{c}$, $\cos\theta = \frac{b}{c}$, and $\tan\theta = \frac{a}{b}$.

    Now try the hypothesis that $\tan\theta = \frac{\sin\theta}{\cos\theta}$.

    That would mean that $\tan\theta = \frac{a}{c} \div \frac{b}{c}$.

    Invert the divisor and multiply: $\tan\theta = \frac{a}{c} \times \frac{c}{b} \rightarrow \rightarrow = \frac{a}{\cancel{c}} \times \frac{\cancel{c}}{b} = \frac{a}{b}$

    How about that??!

2. Prove that $\cot\theta = \frac{\cos\theta}{\sin\theta}$.

    answer: By definition, in right $\triangle ABC$, with the right angle at C and θ at A, $\sin\theta = \frac{a}{c}$, $\cos\theta = \frac{b}{c}$, and $\cot\theta = \frac{b}{a}$.

    Now try the hypothesis that $\cot\theta = \frac{\cos\theta}{\sin\theta}$.

    That would mean that $\cot\theta = \frac{b}{c} \div \frac{a}{c}$.

    Invert the divisor and multiply: $\cot\theta = \frac{b}{c} \times \frac{c}{a} \rightarrow \rightarrow = \frac{b}{\cancel{c}} \times \frac{\cancel{c}}{a} = \frac{b}{a}$

    And there you have it!

3. Given the same right $\triangle ABC$ as in Problems 1 and 2, name and define the reciprocal identities for $\sin\theta$ and $\cos\theta$, in that order.

    answer: $\csc\theta = \frac{1}{\sin\theta} = \frac{c}{a}$, $\sec\theta = \frac{1}{\cos\theta} = \frac{c}{b}$

Cofunction Identities

The cofunction identities may be stated two different ways. The first way is as ratios that demonstrate that the sine of one acute angle in a triangle is the cosine of the other acute angle, and so forth:

$$\sin A = \frac{a}{c} = \cos B \qquad \sin B = \frac{b}{c} = \cos A$$

$$\sec A = \frac{c}{b} = \csc B \qquad \sec B = \frac{c}{a} = \csc A$$

$$\tan A = \frac{a}{b} = \cot B \qquad \tan B = \frac{b}{a} = \cot A$$

The cofunction identities can also be stated in complementary form, since the two acute angles of a right triangle must add up to 90°:

$$\sin\theta = \cos(90° - \theta) \qquad \cos\theta = \sin(90° - \theta)$$
$$\sec\theta = \csc(90° - \theta) \qquad \csc\theta = \sec(90° - \theta)$$
$$\tan\theta = \cot(90° - \theta) \qquad \cot\theta = \tan(90° - \theta)$$

There is really no point in proving the cofunction identities, since you should be able to see them for yourself by just inspecting the following figure.

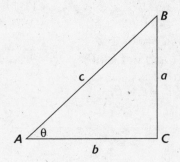

Identities for Negatives

If there is anything simpler than identities for negatives, we don't know what it is; nevertheless, they need to be stated:

$$\sin(-\theta) = -\sin\theta$$
$$\cos(-\theta) = -\cos\theta$$
$$\tan(-\theta) = -\tan\theta$$
$$\sec(-\theta) = -\sec\theta$$
$$\csc(-\theta) = -\csc\theta$$
$$\cot(-\theta) = -\cot\theta$$

Example Problems here would be too silly. Just keep these in the back of your mind, because you never know when this information might come in handy.

Pythagorean Identities

The last group of fundamental identities are the **Pythagorean identities.** Note that each line contains the same identity written in three different forms. If you learn the first formula on each line, you will have access to all of them; however, it pays to recognize the fact that each Pythagorean identity can appear in (**at least**) three different forms:

$$\sin^2\theta + \cos^2\theta = 1 \qquad \sin^2\theta = 1 - \cos^2\theta \qquad \cos^2\theta = 1 - \sin^2\theta$$
$$\csc^2\theta - \cot^2\theta = 1 \qquad 1 + \cot^2\theta = \csc^2\theta \qquad \cot^2\theta = \csc^2\theta - 1$$
$$\sec^2\theta - \tan^2\theta = 1 \qquad 1 + \tan^2\theta = \sec^2\theta \qquad \tan^2\theta = \sec^2\theta - 1$$

Example Problems

These problems show the solutions. All problems refer to the same right triangle as in the previous figure.

1. Prove $\sin^2\theta + \cos^2\theta = 1$.

   answer: In right $\triangle ABC$, $a^2 + b^2 = c^2$.

   Divide by c^2: $\dfrac{a^2}{c^2} + \dfrac{b^2}{c^2} = \dfrac{c^2}{c^2}$

   Rename and simplify: $\left(\dfrac{a}{c}\right)^2 + \left(\dfrac{b}{c}\right)^2 = 1$

   But $\sin\theta = \dfrac{a}{c}$, and $\cos\theta = \dfrac{b}{c}$, so . . .

   You substitute $\left(\dfrac{a}{c}\right)^2 + \left(\dfrac{b}{c}\right)^2 = 1$ and get $(\sin\theta)^2 + (\cos\theta)^2 = 1$.

   But, $(\sin\theta)^2 = \sin^2\theta$ and $(\cos\theta)^2 = \cos^2\theta$.

   Therefore, $\sin^2\theta + \cos^2\theta = 1$.

2. Prove $1 + \tan^2\theta = \sec^2\theta$.

   answer: In right $\triangle ABC$, $a^2 + b^2 = c^2$.

   Divide by b^2: $\dfrac{a^2}{b^2} + \dfrac{b^2}{b^2} = \dfrac{c^2}{b^2}$

   Rename and simplify: $\left(\dfrac{a}{b}\right)^2 + 1 = \left(\dfrac{c}{b}\right)^2$

   But $\tan\theta = \dfrac{a}{b}$, and $\sec\theta = \dfrac{c}{b}$, so . . .

   You substitute $\left(\dfrac{a}{b}\right)^2 + 1 = \left(\dfrac{c}{b}\right)^2$ and get $(\tan\theta)^2 + 1 = (\sec\theta)^2$.

   But, $(\tan\theta)^2 = \tan^2\theta$ and $(\sec\theta)^2 = \sec^2\theta$.

   Therefore, $\tan^2\theta + 1 = \sec^2\theta$, or $1 + \tan^2\theta = \sec^2\theta$.

3.   Prove $1 + \cot^2\theta = \csc^2\theta$.

     answer: In right $\triangle ABC$, $a^2 + b^2 = c^2$.

     Divide by a^2: $\dfrac{a^2}{a^2} + \dfrac{b^2}{a^2} = \dfrac{c^2}{a^2}$

     Rename and simplify: $1 + \left(\dfrac{b}{a}\right)^2 = \left(\dfrac{c}{a}\right)^2$

     But $\cot\theta = \dfrac{b}{a}$, and $\csc\theta = \dfrac{c}{a}$, so . . .

     You substitute $1 + \left(\dfrac{b}{a}\right)^2 = \left(\dfrac{c}{a}\right)^2$ and get $1 + (\cot\theta)^2 = (\csc\theta)^2$.

     But, $(\cot\theta)^2 = \cot^2\theta$ and $(\csc\theta)^2 = \csc^2\theta$.

     Therefore, $1 + \cot^2\theta = \csc^2\theta$.

Work Problems

Use these problems to give yourself additional practice.

For problems 1 and 2, $\sin\phi = \dfrac{5}{6}$ and $\cos\phi < 0$.

1.   Find $\cos\phi$.

2.   Find $\tan\phi$.

3.   Express $\dfrac{\tan\theta + \sec\theta}{1 + \sin\theta}$ in simplest form.

4.   What is the value of $7\sin^2\theta + 7\cos^2\theta$?

Worked Solutions

1. $-\dfrac{\sqrt{11}}{6}$ We need to use an identity that relates sine to cosine and does not require knowledge of an angle.

$$\sin^2\phi + \cos^2\phi = 1$$

Substitute: $\left(\dfrac{5}{6}\right)^2 + \cos^2\phi = 1$

Collect like terms: $\cos^2\phi = 1 - \left(\dfrac{5}{6}\right)^2$

Square the fraction: $\cos^2\phi = 1 - \dfrac{25}{36}$

Subtract: $\cos^2\phi = 1 - \dfrac{25}{36} = \dfrac{11}{36}$

Find square root of each side: $\cos\phi = \dfrac{\sqrt{11}}{6}$

But $\cos\phi < 0$, so $\cos\phi = -\dfrac{\sqrt{11}}{6}$

2. $\dfrac{5\sqrt{11}}{11}$

Since you know the sine and cosine, use the ratio identity:

$$\tan\phi = \frac{\sin\phi}{\cos\phi}$$

Substitute: $\tan\phi = \dfrac{\dfrac{5}{6}}{-\dfrac{\sqrt{11}}{6}}$

Multiply by the divisor's reciprocal (also known as invert and multiply):

$$\tan\phi = \frac{5}{6} \times \left(-\frac{6}{\sqrt{11}}\right)$$

First cancel the sixes: $\tan\phi = \dfrac{5}{\cancel{6}_1} \times \left(-\dfrac{\cancel{6}^{1}}{\sqrt{11}}\right)$

Then multiply: $\tan\phi = \dfrac{-5\sqrt{11}}{11}$

Never leave a radical in the denominator: $\tan\phi = -\dfrac{5}{\sqrt{11}} \cdot \dfrac{\sqrt{11}}{\sqrt{11}} = -\dfrac{5\sqrt{11}}{11}$

Finish up: $\tan\phi = -\dfrac{5\sqrt{11}}{11}$

3. **sec**θ You start out with $\dfrac{\tan\theta + \sec\theta}{1 + \sin\theta}$

Change everything to sin and cos: $\dfrac{\dfrac{\sin\theta}{\cos\theta} + \dfrac{1}{\cos\theta}}{1 + \sin\theta}$

Combine numerators: $\dfrac{\dfrac{\sin\theta + 1}{\cos\theta}}{1 + \sin\theta}$

Invert to make a multiplication problem: $\dfrac{\sin\theta + 1}{\cos\theta} \cdot \dfrac{1}{1 + \sin\theta}$

Cancel and multiply: $\dfrac{\cancel{\sin\theta + 1}}{\cos\theta} \cdot \dfrac{1}{\cancel{1 + \sin\theta}} = \dfrac{1}{\cos\theta}$

Recognize the secant's reciprocal: $\dfrac{1}{\cos\theta} = \sec\theta$

4. **7**   Does the expression $7\sin^2\theta + 7\cos^2\theta$ look familiar?

You know that $\sin^2\theta + \cos^2\theta = 1$.

Multiply the equation by 7: $7(\sin^2\theta + \cos^2\theta = 1)$.

That will give you $7\sin^2\theta + 7\cos^2\theta = 7$.

And there's your answer; the expression equals 7.

Addition and Subtraction Identities

The fundamental identities discussed previously all involved working with a single variable. The following identities utilize two variables and are known as **addition and subtraction identities, sum and difference identities,** or simply **trigonometric addition identities** for sine and cosine.

1. $\sin(\alpha + \beta) = \sin\alpha\cos\beta + \cos\alpha\sin\beta$ The sum identity for sine.

2. $\sin(\alpha - \beta) = \sin\alpha\cos\beta - \cos\alpha\sin\beta$ The difference identity for sine.

3. $\cos(\alpha + \beta) = \cos\alpha\cos\beta - \sin\alpha\sin\beta$ The sum identity for cosine.

4. $\cos(\alpha - \beta) = \cos\alpha\cos\beta + \sin\alpha\sin\beta$ The difference identity for cosine.

Example Problems

These problems show the answers and solutions.

1. Change $\sin 70°\cos 120° + \cos 70°\sin 120°$ into a trigonometric function in a single variable and evaluate it.

   answer: −.174

   This is one side of the sum identity for sines (1, above): $\sin(\alpha + \beta) = \sin\alpha\cos\beta + \cos\alpha\sin\beta$

   Complete the sum identity
   for sine: $\sin 70°\cos 120° + \cos 70°\sin 120° = \sin(70° + 120°)$

   Add inside the parentheses: $= \sin(190°)$

   This is a third quadrant angle, $[180° < 190° < 270°]$,
   so sin is negative: $= -\sin 190°$

   To get the reference angle in quadrant III, subtract 180°: $= -\sin(190° - 180°$

   Subtract: $= -\sin 10°$

        $= -.174$

2. Change $\cos 130°\cos 70° + \sin 130°\sin 70°$ into a trigonometric function in a single variable and evaluate it.

   answer: 0.5 or $\frac{1}{2}$

   This is part of the difference identity for cosines (4, above):

        $\cos(\alpha - \beta) = \cos\alpha\cos\beta + \sin\alpha\sin\beta$

   Complete the difference
   identity for cosine: $\cos 130°\cos 70° + \sin 130°\sin 70° = \cos(130° - 70°)$

   Subtract inside the parentheses: $= \cos 60°$

   All functions are positive in the first quadrant, so evaluate: $\cos 60° = 0.5$

3. Change $\cos130°\cos70° - \sin130°\sin70°$ into a trigonometric function in a single variable and evaluate it.

    answer: −0.94

    This is part of the sum identity for cosines (3, above): $\cos(\alpha + \beta) = \cos\alpha\cos\beta - \sin\alpha\sin\beta$

    Complete the sum identity for
    cosine: $\cos130°\cos70° - \sin130°\sin70° = \cos(130° + 70°)$

    Add inside the parentheses: $= \cos(200°)$

    This is a third quadrant angle, $[180° < 200° < 270°]$,
    so cos is negative: $= -\cos200°$

    To get the reference angle in quadrant III, subtract 180°: $= -\cos(200° - 180°)$

    Subtract: $= -\cos20°$

    Evaluate: $\cos20° = -0.94$

Double Angle Identities

The double angle identities for sine and cosine, as well as the half angle identities, are derived from the sum and difference identities given previously. They are shown here:

$$\sin2\theta = 2\sin\theta\cos\theta$$
$$\cos2\theta = \cos^2\theta - \sin^2\theta$$
$$\cos2\theta = 2\cos^2\theta - 1$$
$$\cos2\theta = 1 - 2\sin^2\theta$$

These identities are used to express the sine or cosine of an angle that's double the size of the one you know, in terms of the sine and cosine of that known angle. You'll notice that there are three times as many forms with cosine to the left of the equal sign as there are for sine in that location.

Example Problems

These problems show the answers and solutions.

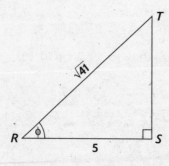

1. If $\cos\phi = \dfrac{5}{\sqrt{41}}$, find the value of $\sin2\phi$.

    answer: $\dfrac{40}{41}$

Start out with the Pythagorean theorem: $r^2 = \left(\sqrt{41}\right)^2 - 5^2$

Solving that, you first square the numerics: $r^2 = 41 - 25$

Subtract: $r^2 = 16$

Take the square root of both sides: $r = 4$

Since $\sin = \dfrac{\text{opposite}}{\text{hypotenuse}}$, $\sin \phi = \dfrac{4}{\sqrt{41}}$.

Now start with the identity: $\sin 2\phi = 2\sin\phi\cos\phi$

Substitute what you know: $\sin 2\phi = 2\left(\dfrac{4}{\sqrt{41}}\right)\left(\dfrac{5}{\sqrt{41}}\right)$

Next group to multiply: $\sin 2\phi = \dfrac{2\cdot 4\cdot 5}{\sqrt{41}\cdot\sqrt{41}}$

Finally, multiply: $\sin 2\phi = \dfrac{40}{41}$

2. If P is in the second quadrant and $\sin P = \dfrac{3}{5}$, find $\sin 2P$.

answer: $\dfrac{24}{25}$

If $\sin P = \dfrac{3}{5}$, then $\cos P = \dfrac{4}{5}$. (Remember the 3-4-5 right triangle? Otherwise, use the Pythagorean theorem to find the adjacent side.)

Now start with the identity: $\sin 2P = 2\sin P\cos P$

Substitute what you know: $\sin 2P = 2\left(\dfrac{3}{5}\right)\left(\dfrac{4}{5}\right)$

Next group to multiply: $\sin 2P = \dfrac{2\cdot 3\cdot 4}{5\cdot 5}$

Finally, multiply: $\sin 2P = \dfrac{24}{25}$

Remember that sine is positive in the second quadrant, so you need to do nothing further.

3. If A is a second quadrant angle, and $\sin A = \dfrac{5}{13}$, what is $\cos 2A$?

answer: $-\dfrac{119}{169}$

For openers, you might use either the Pythagorean theorem or your knowledge of Pythagorean triples to find the third side of the triangle, 12. Therefore, $\cos A = \dfrac{12}{13}$.

Of course, you don't need that information if you use the identity:

$$\cos 2\theta = 1 - 2\sin^2\theta$$

Substitute: $\cos 2\theta = 1 - 2\left(\dfrac{5}{13}\right)^2$

Clear the parentheses: $\cos 2\theta = 1 - \dfrac{2}{1}\cdot\dfrac{25}{169}$

Multiply: $\cos 2\theta = 1 - \dfrac{50}{169}$

Subtract: $\cos 2\theta = \dfrac{169}{169} - \dfrac{50}{169} = \dfrac{119}{169}$

And since cosine is negative in quadrant II: $\cos 2\theta = -\dfrac{119}{169}$

Half Angle Identities

If the double angle identities seemed strange to you, the **half angle identities** should seem even more so. These identities are used to express the sine or cosine of an angle that's half the size of the one you know, in terms of the sine and cosine of that known angle. (By the way, don't think there are no double or half angle identities for tangent. We're just putting them off for a while.) The sine and cosine half angle identities are as follows:

$$\sin\frac{\theta}{2} = \pm\sqrt{\frac{1-\cos\theta}{2}}$$
$$\cos\frac{\theta}{2} = \pm\sqrt{\frac{1+\cos\theta}{2}}$$

Notice that each radical sign is preceded by a $\pm$ (which means plus or minus). That's because the sign of the answer will depend upon which quadrant the half angle falls in. If $\theta = 280°$, then $\frac{\theta}{2} = 140°$. Since $140°$ is a quadrant II angle, $\sin\theta$ would be preceded by a positive sign, and $\cos\theta$ would be preceded by a negative.

Example Problems

These problems show the answers and solutions.

1. If $\cos\theta = -\dfrac{5}{16}$ and θ is in quadrant III, find $\cos\frac{1}{2}\theta$.

    answer: $-\dfrac{\sqrt{22}}{8}$

    If θ is in quadrant III, then θ is between $270°$ and $360°$. That means $\frac{1}{2}\theta$ is greater than $135°$ and less than $180°$. That's in quadrant II. In quadrant II, cosine is also negative.

    Now write the relevant equation: $\cos\dfrac{\theta}{2} = \pm\sqrt{\dfrac{1+\cos\theta}{2}}$

    Substitiute and adjust the sign: $\cos\dfrac{\theta}{2} = -\sqrt{\dfrac{1+\left(-\dfrac{5}{16}\right)}{2}}$

    Express both numbers to be added in like fractions: $\cos\dfrac{\theta}{2} = -\sqrt{\dfrac{\dfrac{16}{16}+\left(-\dfrac{5}{16}\right)}{2}}$

    Combine fractions: $\cos\dfrac{\theta}{2} = -\sqrt{\dfrac{\dfrac{11}{16}}{2}}$

    Divide: $\cos\dfrac{\theta}{2} = -\sqrt{\dfrac{11}{16}\div\dfrac{2}{1}} = -\sqrt{\dfrac{11}{16}\times\dfrac{1}{2}} = -\sqrt{\dfrac{11}{32}}$

Work out what's under the radical sign: $\cos\dfrac{\theta}{2} = -\dfrac{\sqrt{11}}{4\sqrt{2}}$

Rationalize the denominator: $\cos\dfrac{\theta}{2} = -\dfrac{\sqrt{11}}{4\sqrt{2}} \times \dfrac{\sqrt{2}}{\sqrt{2}} = -\dfrac{\sqrt{22}}{8}$

2. If ϕ is a positive acute angle and $\cos\phi = \dfrac{1}{9}$, find $\sin\dfrac{\phi}{2}$.

   answer: $\dfrac{2}{3}$

   First write the formula for the half angle sine identity: $\sin\dfrac{\phi}{2} = \pm\sqrt{\dfrac{1 - \cos\phi}{2}}$

   Since ϕ is a positive acute angle, sine is positive: $\sin\dfrac{\phi}{2} = \sqrt{\dfrac{1 - \cos\phi}{2}}$

   Next, substitute: $\sin\dfrac{\phi}{2} = \sqrt{\dfrac{1 - \dfrac{1}{9}}{2}}$

   Subtract: $\sin\dfrac{\phi}{2} = \sqrt{\dfrac{\dfrac{9}{9} - \dfrac{1}{9}}{2}} = \sqrt{\dfrac{\dfrac{8}{9}}{2}}$

   Divide by 2: $\sin\dfrac{\phi}{2} = \sqrt{\dfrac{8}{9} \div \dfrac{2}{1}} = \sqrt{\dfrac{8}{9} \times \dfrac{1}{2}} = \sqrt{\dfrac{4}{9}}$

   Work out what's under the radical sign: $\sin\dfrac{\phi}{2} = \dfrac{2}{3}$

3. If $\sin\dfrac{1}{2}\theta = \dfrac{1}{4}$ and θ is an acute angle, find $\cos\theta$.

   answer: $\dfrac{7}{8}$

   The only identity you have that relates the two functions is: $\sin\dfrac{\theta}{2} = \pm\sqrt{\dfrac{1 - \cos\theta}{2}}$

   Substitute and make radical positive, since angle is acute: $\dfrac{1}{4} = \sqrt{\dfrac{1 - \cos\theta}{2}}$

   Square both sides of the equation: $\left(\dfrac{1}{4} = \sqrt{\dfrac{1 - \cos\theta}{2}}\right)^2 \rightarrow\rightarrow \dfrac{1}{16} = \dfrac{1 - \cos\theta}{2}$

   Next, cross multiply: $16 - 16\cos\theta = 2$

   Subtract 16 from both sides: $-16\cos\theta = 2 - 16 = -14$

   Divide both sides by -16; simplify: $\cos\theta = \dfrac{-14}{-16} = \dfrac{7}{8}$

Work Problems

Use these problems to give yourself additional practice.

1. Change $\sin 50°\cos 100° + \cos 50°\sin 100°$ into a trigonometric function in a single variable and evaluate it.

2. Change $\cos 140°\cos 60° + \sin 140°\sin 60°$ into a trigonometric function in a single variable and evaluate it.

3. If $\cos\phi = 0.8$, find the value of $\sin 2\phi$.

4. If A is a second quadrant angle, and $\sin A = \dfrac{9}{15}$, what is $\cos 2A$?

5. If $\cos\theta = -\dfrac{3}{8}$ and θ is in quadrant III, find $\cos \dfrac{1}{2}\theta$.

Worked Solutions

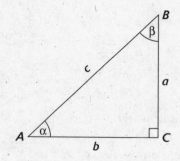

1. **0.5 or $\dfrac{1}{2}$**   This is one side of the sum identity for sines: $\sin(\alpha + \beta) = \sin\alpha\cos\beta + \cos\alpha\sin\beta$

   Complete the sum identity for sine: $\sin 50°\cos 100° + \cos 50°\sin 100° = \sin(50° + 100°)$

   Add inside the parentheses: $= \sin(150°)$

   This is a second quadrant angle, $[90° < 150° < 180°]$,
   so sin is positive: $= \sin 150°$

   To get the reference angle in quadrant II, subtract from $180°$: $= \sin(180° - 150°)$

   Subtract: $= \sin 30°$

                                                                                                  $= 0.5$

2. **0.174**   This is part of the difference identity for cosines: $\cos(\alpha - \beta) = \cos\alpha\cos\beta + \sin\alpha\sin\beta$.

   Complete the difference identity for cosine:

   $$\cos 140°\cos 60° + \sin 140°\sin 60° = \cos(140° - 60°)$$

   Subtract inside the parentheses: $= \cos 80°$

   All functions are positive in the first quadrant, so evaluate: $\cos 80° = 0.174$

3. **0.96** 0.8 may also be written as $\frac{8}{10}$, which gives us two sides of the triangle.

   To get the third, use the Pythagorean theorem: $r^2 = 10^2 - 8^2$

   Solving that, you first square the numerics: $r^2 = 100 - 64$

   Subtract: $r^2 = 36$

   Take the square root of both sides: $r = 6$

   Since $\sin = \frac{\text{opposite}}{\text{hypotenuse}}$, $\sin\phi = \frac{6}{10}$.

   Now start with the identity: $\sin 2\phi = 2\sin\phi\cos\phi$

   Substitute what you know: $\sin 2\phi = 2(0.8)(0.6)$.

   Finally, multiply: $\sin 2\phi = 0.96$

4. $-\frac{63}{225}$ For openers, you might use either the Pythagorean Theorem or your knowledge of Pythagorean triples to find the third side of the triangle, 12. Therefore, the $\cos A = \frac{12}{15}$.

   Of course, you don't need that information if you use the identity:

   $$\cos 2\theta = 1 - 2\sin^2\theta$$

   Substitute: $\cos 2\theta = 1 - 2\left(\frac{9}{15}\right)^2$

   Clear the parentheses: $\cos 2\theta = 1 - \frac{2}{1} \cdot \frac{81}{225}$

   Multiply: $\cos 2\theta = 1 - \frac{162}{225}$

   Subtract: $\cos 2\theta = \frac{225}{225} - \frac{162}{225} = \frac{63}{225}$

   And since cosine is negative in quadrant II: $\cos 2\theta = -\frac{63}{225}$

5. $-\frac{\sqrt{5}}{4}$ If θ is in quadrant III, then θ is between 270° and 360°. That means $\frac{1}{2}\theta$ is greater than 135° and less than 180°. That's in quadrant II. In quadrant II, cosine is also negative.

   Now write the relevant equation: $\cos\frac{\theta}{2} = \pm\sqrt{\frac{1 + \cos\theta}{2}}$

   Substitiute and adjust the sign: $\cos\frac{\theta}{2} = -\sqrt{\frac{1 + \left(-\frac{3}{8}\right)}{2}}$

   Express both numbers to be added in like fractions: $\cos\frac{\theta}{2} = -\sqrt{\frac{\frac{8}{8} + \left(-\frac{3}{8}\right)}{2}}$

   Combine fractions: $\cos\frac{\theta}{2} = -\sqrt{\frac{\frac{5}{8}}{2}}$

   Divide: $\cos\frac{\theta}{2} = -\sqrt{\frac{5}{8} \div \frac{2}{1}} = -\sqrt{\frac{5}{8} \times \frac{1}{2}} = -\sqrt{\frac{5}{16}}$

   Work out what's under the radical sign: $\cos\frac{\theta}{2} = -\frac{\sqrt{5}}{4}$

Tangent Identities

The tangent identities are together near the end of this chapter, because they're all derived from the sine and cosine identities using the ratio identitiy $\tan \theta = \dfrac{\sin \theta}{\cos \theta}$ with the several formulas you've studied in this chapter. The **tangent sum and difference identities** look like this:

$$\tan(\alpha + \beta) = \frac{\tan \alpha + \tan \beta}{1 - \tan \alpha \tan \beta}$$

$$\tan(\alpha - \beta) = \frac{\tan \alpha - \tan \beta}{1 + \tan \alpha \tan \beta}$$

The **double angle identity for tangent** is derived using the tangent sum identity, and if you really want to see how it's done, you can check *Cliffs Quick Review Trigonometry*. The formula looks like this:

$$\tan 2\theta = \frac{2 \tan \theta}{1 - \tan^2 \theta}$$

The **tangent half angle identity** has three different forms. In the last of those forms, its sign is determined by the quadrant in which the angle terminates.

$$\tan \frac{\theta}{2} = \frac{1 - \cos \theta}{\sin \theta}$$

$$\tan \frac{\theta}{2} = \frac{\sin \theta}{1 + \cos \theta}$$

$$\tan \frac{\theta}{2} = \theta \pm \sqrt{\frac{1 - \cos \theta}{1 + \cos \theta}}$$

Example Problems

These problems show the answers and solutions.

1. If $\tan \alpha = 3$ and $\tan \beta = \dfrac{1}{2}$, what is the tangent of their sum?

   answer: –7

   $$\tan(\alpha + \beta) = \frac{\tan \alpha + \tan \beta}{1 - \tan \alpha \tan \beta}$$

   Substitute: $\tan(\alpha + \beta) = \dfrac{3 + \dfrac{1}{2}}{1 - \left(3 \times \dfrac{1}{2}\right)}$

   Combine terms: $\tan(\alpha + \beta) = \dfrac{\dfrac{6}{2} + \dfrac{1}{2}}{1 - \left(\dfrac{3}{2}\right)} = \dfrac{\dfrac{7}{2}}{-\dfrac{1}{2}}$

   Divide: $\tan(\alpha + \beta) = \dfrac{7}{2} \div -\dfrac{1}{2} = \dfrac{7}{2} \times -\dfrac{2}{1} = -7$

2. If $\tan \alpha = 2$ and $\tan \beta = \frac{1}{2}$, what is the tangent of their difference $(\alpha - \beta)$?

answer: $\frac{3}{4}$

First, state the formula: $\tan(\alpha - \beta) = \dfrac{\tan\alpha - \tan\beta}{1 + \tan\alpha\tan\beta}$

Substitute: $\tan(\alpha - \beta) = \dfrac{2 - \frac{1}{2}}{1 + \left(2 \times \frac{1}{2}\right)}$

Combine terms: $\tan(\alpha - \beta) = \dfrac{\frac{4}{2} - \frac{1}{2}}{1 + \left(\frac{4}{2} \times \frac{1}{2}\right)} = \dfrac{\frac{3}{2}}{1 + \frac{4}{4}} = \dfrac{\frac{3}{2}}{\frac{2}{1}}$

And divide: $\tan(\alpha - \beta) = \dfrac{3}{2} \div \dfrac{2}{1} = \dfrac{3}{2} \times \dfrac{1}{2} = \dfrac{3}{4}$

3. If $\theta = \arcsin \frac{3}{5}$, find $\tan 2\theta$.

answer: $\frac{24}{7}$ or $3\frac{3}{7}$ or **3.429**

$\theta = \arcsin \frac{3}{5}$ means that θ is the main value of the angle whose $\sin = \frac{3}{5}$. From the Pythagorean triple, you know that the adjacent side to $\angle\theta$ must be 4, so $\tan \theta = \frac{3}{4}$.

Write the formula: $\tan 2\theta = \dfrac{2\tan\theta}{1 - \tan^2\theta}$

Substitute: $\tan 2\theta = \dfrac{2\left(\frac{3}{4}\right)}{1 - \left(\frac{3}{4}\right)^2}$

Simplify numerator and denominator: $\tan 2\theta = \dfrac{\frac{6}{4}}{\frac{16}{16} - \frac{9}{16}} = \dfrac{\frac{6}{4}}{\frac{7}{16}}$

Finally, divide: $\tan 2\theta = \dfrac{6}{4} \div \dfrac{7}{16} = \dfrac{6}{\cancel{4}_1} \times \dfrac{\cancel{16}^4}{7} = \dfrac{24}{7}$

4. If the cosine of a third quadrant angle is $-\frac{2}{3}$, find the tangent of an angle half that size.

answer: $\sqrt{5}$

Select the half angle tangent formula that applies: $\tan\dfrac{\theta}{2} = \pm\sqrt{\dfrac{1 - \cos\theta}{1 + \cos\theta}}$

Realize that the cosine was negative because it was a quadrant III angle. Tangent in that quadrant is positive.

Substitute and make positive: $\tan\dfrac{\theta}{2} = \sqrt{\dfrac{1 - \left(-\frac{2}{3}\right)}{1 + \left(-\frac{2}{3}\right)}}$

Simplify numerator and denominator: $\tan\dfrac{\theta}{2} = \sqrt{\dfrac{1-\left(-\frac{2}{3}\right)}{1+\left(-\frac{2}{3}\right)}} = \sqrt{\dfrac{\frac{5}{3}}{\frac{1}{3}}}$

Finally, divide: $\tan\dfrac{\theta}{2} = \sqrt{\dfrac{5}{3}\div\dfrac{1}{3}} = \sqrt{\dfrac{5}{3}\times\dfrac{3}{1}} = \sqrt{5}$

Product-Sum and Sum-Product Identities

Sometimes the difference between finding an easy solution to a trigonometric problem and no solution at all is the ability to convert a multiplication to an addition problem or an addition to a multiplication problem. Two sets of identities were developed for doing just this.

Product-Sum Identities

This group of identities is known as **product-sum identities.** They get that name because multiplication, which produces a product, comes on the left, and addition, which produces a sum, is on the right.

$$\sin\alpha\cos\beta = \frac{1}{2}\left[\sin(\alpha+\beta)+\sin(\alpha-\beta)\right]$$

$$\cos\alpha\sin\beta = \frac{1}{2}\left[\sin(\alpha+\beta)-\sin(\alpha-\beta)\right]$$

$$\sin\alpha\sin\beta = \frac{1}{2}\left[\cos(\alpha-\beta)-\cos(\alpha+\beta)\right]$$

$$\cos\alpha\cos\beta = \frac{1}{2}\left[\cos(\alpha+\beta)+\cos(\alpha-\beta)\right]$$

Example Problems

These problems show the answers and solutions.

1. Write cos3c cos2c as a sum.

   answer: $\dfrac{\cos 5c}{2} + \dfrac{\cos c}{2}$

   Pick the appropriate formula: $\cos\alpha\cos\beta = \frac{1}{2}\left[\cos(\alpha+\beta)+\cos(\alpha-\beta)\right]$

   Substitute: $\cos 3c\cos 2c = \frac{1}{2}\left[\cos(3c+2c)+\cos(3c-2c)\right]$

   Clear parentheses: $\cos 3c\cos 2c = \frac{1}{2}\left[\cos 5c+\cos c\right]$

   Clear the brackets: $\cos 3c\cos 2c = \dfrac{\cos 5c}{2} + \dfrac{\cos c}{2}$

2. Express sin5d cos3d as a sum.

   answer: **sin4d + sind**

Pick the appropriate formula:	$\sin\alpha\cos\beta = \frac{1}{2}\left[\sin(\alpha+\beta) + \sin(\alpha-\beta)\right]$
Substitute:	$\sin\alpha\cos\beta = \frac{1}{2}\left[\sin(5d+3d) + \sin(5d-3d)\right]$
Combine terms in parentheses:	$\sin\alpha\cos\beta = \frac{1}{2}\left[\sin8d + \sin2d\right]$
Clear the brackets:	$\sin\alpha\cos\beta = \frac{\sin8d}{2} + \frac{\sin2d}{2}$
Simplify:	$\sin\alpha\cos\beta = \sin4d + \sin d.$

Sum-Product Identities

This group of identities is known as **sum-product identities** because sums are on the left and products are on the right.

$$\sin\alpha + \sin\beta = 2\sin\frac{\alpha+\beta}{2}\cos\frac{\alpha-\beta}{2}$$

$$\sin\alpha - \sin\beta = 2\cos\frac{\alpha+\beta}{2}\sin\frac{\alpha-\beta}{2}$$

$$\cos\alpha + \cos\beta = 2\cos\frac{\alpha+\beta}{2}\cos\frac{\alpha-\beta}{2}$$

$$\cos\alpha - \cos\beta = -2\sin\frac{\alpha+\beta}{2}\sin\frac{\alpha-\beta}{2}$$

Example Problems

These problems show the answers and solutions.

1. Write the sum $\sin4w + \sin6w$ as a product.

   answer: 2sin5wcos − w, or −2 sin5wcosw.

Select the appropriate formula:	$\sin\alpha + \sin\beta = 2\sin\frac{\alpha+\beta}{2}\cos\frac{\alpha-\beta}{2}$
Substitute:	$\sin4w + \sin6w = 2\sin\frac{4w+6w}{2}\cos\frac{4w-6w}{2}$
Combine terms:	$\sin4w + \sin6w = 2\sin\frac{10w}{2}\cos\frac{-2w}{2}$
Divide:	$\sin4w + \sin6w = 2\sin5w\cos - w$ or $-2\sin5\cos w$

2. Write the difference $\cos75° - \cos45°$ as a product.

   answer: −2sin60°sin15°

Select the appropriate formula:	$\cos\alpha - \cos\beta = -2\sin\frac{\alpha+\beta}{2}\sin\frac{\alpha-\beta}{2}$
Substitute:	$\cos75° - \cos45° = -2\sin\frac{75°+45°}{2}\sin\frac{75°-45°}{2}$
Combine terms:	$\cos75° - \cos45° = -2\sin\frac{120°}{2}\sin\frac{30°}{2}$
Divide:	$\cos75° - \cos45° = -2\sin60°\sin15°$

Work Problems

Use these problems to give yourself additional practice.

1. If $\tan\alpha = 4$ and $\tan\beta = \frac{1}{2}$, what is the tangent of their sum?

2. If $\tan\alpha = 3$ and $\tan\beta = \frac{1}{4}$, what is the tangent of their difference $(\alpha - \beta)$?

3. Express $\cos 7a \sin 3a$ as a sum.

4. Express $\sin 6b \sin 8b$ as a sum.

5. Write the sum $\cos 5u + \cos 9u$ as a product.

6. Write the difference $\sin 75° - \sin 45°$ as a product.

Worked Solutions

1. $-\frac{9}{2}$ Write the formula: $\tan(\alpha + \beta) = \dfrac{\tan\alpha + \tan\beta}{1 - \tan\alpha\tan\beta}$

   Substitute: $\tan(\alpha + \beta) = \dfrac{4 + \frac{1}{2}}{1 - \left(4 \times \frac{1}{2}\right)}$

   Combine terms: $\tan(\alpha + \beta) = \dfrac{\frac{9}{2}}{1 - (2)} = \dfrac{\frac{9}{2}}{-1}$

   Divide: $\tan(\alpha + \beta) = -\dfrac{9}{2}$

2. $\frac{11}{7}$ First state the formula: $\tan(\alpha - \beta) = \dfrac{\tan\alpha - \tan\beta}{1 + \tan\alpha\tan\beta}$

   Substitute: $\tan(\alpha - \beta) = \dfrac{3 - \frac{1}{4}}{1 + 3 \times \frac{1}{4}}$

   Combine terms: $\tan(\alpha - \beta) = \dfrac{\frac{11}{4}}{\frac{7}{4}}$

   And divide: $\tan(\alpha - \beta) = \dfrac{11}{4} \div \dfrac{7}{4} = \dfrac{11}{\cancel{4}1} \times \dfrac{\cancel{4}1}{7} = \dfrac{11}{7}$

3. **$\sin 5\alpha - \sin 2\alpha$**

   Pick the appropriate formula: $\cos\alpha \sin\beta = \dfrac{1}{2}\left[\sin(\alpha + \beta) - \sin(\alpha - \beta)\right]$

   Substitute: $\cos 7a\sin 3a = \dfrac{1}{2}\left[\sin(7a + 3a) - \sin(7a - 3a)\right]$

   Combine terms in parentheses: $\cos 7a\sin 3a = \dfrac{1}{2}\left[\sin 10a - \sin 4a\right]$

   Clear the brackets: $\cos 7a\sin 3a = \dfrac{\sin 10a}{2} - \dfrac{\sin 4a}{2}$

   Simplify: $\cos 7a\sin 3a = \sin 5a - \sin 2a$

4.   **−cos*b* −cos7*b***

Pick the appropriate formula: $\sin\alpha\sin\beta = \frac{1}{2}\left[\cos(\alpha-\beta)-\cos(\alpha+\beta)\right]$

Substitute: $\sin 6b\sin 8b = \frac{1}{2}\left[\cos(6b-8b)-\cos(6b+8b)\right]$

Combine terms in parentheses: $\sin 6b\sin 8b = \frac{1}{2}\left[\cos - 2b - \cos 14b\right]$

Clear the brackets: $\sin 6b\sin 8b = \frac{\cos - 2b}{2} - \frac{\cos 14b}{2}$

Simplify: $\sin 6b\sin 8b = -\cos b - \cos 7b$

5.   **2cos7*u*cos − 2*u* or −2 cos7*u*cos 2*u***

Select the appropriate formula: $\cos\alpha + \cos\beta = 2\cos\frac{\alpha+\beta}{2}\cos\frac{\alpha-\beta}{2}$

Substitute: $\cos 5u + \cos 9u = 2\cos\frac{5u+9u}{2}\cos\frac{5u-9u}{2}$

Combine terms: $\cos 5u + \cos 9u = 2\cos\frac{14u}{2}\cos\frac{-4u}{2}$

Divide: $\cos 5u + \cos 9u = 2\cos 7u\cos - 2u$ or $-2\cos 7\cos 2u$

6.   **2cos60°sin15°**

Select the appropriate formula: $\sin\alpha - \sin\beta = 2\cos\frac{\alpha+\beta}{2}\sin\frac{\alpha-\beta}{2}$

Substitute: $\sin 75° - \sin 45° = 2\cos\frac{75°+45°}{2}\sin\frac{75°-45°}{2}$

Combine terms: $\sin 75° - \sin 45° = 2\cos\frac{120°}{2}\sin\frac{30°}{2}$

Divide: $\sin 75° - \sin 45° = 2\cos 60°\sin 15°$

Chapter Problems and Solutions

Problems

Solve these problems for more practice applying the skills from this chapter. Worked out solutions follow problems.

1.   Write the reciprocal identity for cosecant.

2.   Write the ratio identity for cotangent.

3.   Write the cofunction identity for secant in both ratio and complementary form.

4.   Write an expression that is always equivalent to csc(−θ).

5.   What is the value of $11\sin^2\theta + 11\cos^2\theta$?

6.   Rewrite the expression in simplest form: $\cot\phi(\sin\phi - \sec\phi)$.

7. Rewrite the expression in simplest form: $\dfrac{\tan^2\theta}{\tan^2\theta + 1}$.

8. Change $\cos120°\cos40° + \sin120°\sin40°$ into a trigonometric function in a single variable and evaluate it.

9. Change $\sin70°\cos100° + \cos70°\sin100°$ into a trigonometric function in a single variable and evaluate it.

10. If $\cos\phi = \dfrac{9}{3\sqrt{13}}$, find the value of $\sin2\phi$.

11. If P is in the second quadrant and $\sin P = \dfrac{8}{10}$, find $\sin2P$.

12. If A is a second quadrant angle, and $\sin A = \dfrac{12}{13}$, what is $\cos 2A$?

13. If ϕ is a positive acute angle and $\cos\phi = \dfrac{3}{7}$, find $\sin\dfrac{\phi}{2}$.

14. If $\sin\dfrac{1}{2}\phi = \dfrac{2}{5}$ and ϕ is an acute angle, find $\cos\phi$.

15. If $\cos\theta = -\dfrac{3}{8}$ and θ is in quadrant III, find $\cos\dfrac{1}{2}\theta$.

16. If $\tan\alpha = 4$ and $\tan\beta = \dfrac{1}{3}$, what is the tangent of their sum?

17. If $\tan\alpha = 2$ and $\tan\beta = \dfrac{1}{4}$, what is the tangent of their difference $(\alpha - \beta)$?

18. If $\theta = \arcsin\dfrac{4}{5}$, find $\tan 2\theta$.

19. If the cosine of a third quadrant angle is $-\dfrac{3}{4}$, find the tangent of an angle half that size.

20. Write $\cos4w\cos5w$ as a sum.

21. Express $\sin6d\cos2d$ as a sum.

22. Express $\cos5a\sin4a$ as a sum.

23. Write the difference $\cos60° - \cos50°$ as a product.

24. Write the sum $\sin8u + \sin 5u$ as a product.

25. Write the difference $\sin55° - \sin45°$ as a product.

Answers and Solutions

1. ***Answer:*** $\dfrac{1}{\sin\theta}$ $\csc\theta = \dfrac{1}{\sin\theta}$ is the reciprocal identity in question. You studied reciprocal identities in Chapter 1.

2. ***Answer:*** $\cot\theta = \dfrac{\cos\theta}{\sin\theta}$ The ratio identities express tangent and cotangent in terms of sine and cosine.

3. ***Answer:*** $\sec A = \frac{c}{b} = \csc B$, $\sec\theta = \csc(90° - \theta)$ $\sec A = \frac{c}{b} = \csc B$ is the ratio form of the cofunction identity for secant; $\sec\theta = \csc(90° - \theta)$ is the complementary form.

4. ***Answer:*** $-\csc\theta$ $\csc(-\theta) = -\csc\theta$ is the negative identity for $\csc(-\theta)$.

5. ***Answer:*** 11   Does the expression $11\sin^2\theta + 11\cos^2\theta$ look familiar?

   You have learned the Pythagorean identity, $\sin^2\theta + \cos^2\theta = 1$.

   So, multiply the equation by 11: $11(\sin^2\theta + \cos^2\theta = 1)$

   That yields: $11\sin^2\theta + 11\cos^2\theta = 11$

   So, the expression equals 11

6. ***Answer:*** $\cos\phi - \csc\phi$ Rewrite in terms of sin and cos: $\frac{\cos\phi}{\sin\phi}\left(\sin\phi - \frac{1}{\cos\phi}\right)$

   Clear parentheses: $\frac{\cos\phi\sin\phi}{\sin\phi} - \frac{\cos\phi}{\cos\phi\sin\phi}$

   So, a bit of cancelling: $\frac{\cos\phi\sin\phi}{\sin\phi} - \frac{\cos\phi}{\cos\phi\sin\phi} = \cos\phi - \frac{1}{\sin\phi}$

   And there you have it: $\cos\phi - \csc\phi$

7. ***Answer:*** $\sin2\theta$

   State the expression: $\frac{\tan^2\theta}{\tan^2\theta + 1}$

   Substitute for $\tan^2$ in both terms: $\frac{\frac{\sin^2\theta}{\cos^2\theta}}{(\sec^2\theta - 1) + 1}$

   Add 1 and -1 and sub sec's identification: $\frac{\frac{\sin^2\theta}{\cos^2\theta}}{\frac{1}{\cos^2\theta}}$

   Divide: $\frac{\sin^2\theta}{\cos^2\theta} \div \frac{1}{\cos^2\theta} = \frac{\sin^2\theta}{\cos^2\theta} \cdot \frac{\cos^2\theta}{1} = \sin^2\theta$

8. ***Answer:*** 0.174

   $\cos120°\cos40° + \sin120°\sin40°$ is part of the difference identity for cosines: $\cos(\alpha - \beta) = \cos\alpha\cos\beta + \sin\alpha\sin\beta$.

   Complete the difference identity for cosine:
   $\cos120°\cos40° + \sin120°\sin40° = \cos(120° - 40°)$

   Subtract inside the parentheses: $= \cos80°$

   All functions are positive in the first quadrant, so evaluate: $\cos80° = 0.174$

9. **Answer:** 0.174

    sin70°cos100° + cos70°sin100° is one
    side of the sum identity for sines: $\qquad$ $\sin(\alpha + \beta) = \sin\alpha\cos\beta + \cos\alpha\sin\beta.$

    Complete the sum identity
    for sine: $\qquad$ $\sin70°\cos100° + \cos70°\sin100° = \sin(70° + 100°)$

    Add inside the parentheses: $\qquad$ $= \sin(170°)$

    This is a second quadrant angle, [90° < 170° < 180°],
    so sin is positive: $\qquad$ $= \sin170°$

    To get the reference angle in quadrant II, subtract from 180°: $\qquad$ $= \sin(180° - 170°)$

    Subtract: $\qquad$ $= \sin10°$

    $\qquad$ $= 0.174$

10. **Answer:** $\dfrac{12}{13}$

    Start out with the Pythagorean theorem: $\qquad$ $r^2 = \left(3\sqrt{13}\right)^2 - 9^2$

    Solving that, you first square the numerics: $\qquad$ $r^2 = 9 \cdot 13 - 81 = 117 - 81$

    Subtract: $\qquad$ $r^2 = 36$

    Take the square root of both sides: $\qquad$ $r = 6$

    Since $\dfrac{\text{opposite}}{\text{hypotenuse}}$, $\qquad$ $\sin\phi = \dfrac{6}{3\sqrt{13}}$

    Now start with the identity: $\qquad$ $\sin2\phi = 2\sin\phi\cos\phi$

    Substitute, note $3\sqrt{13} = \sqrt{9 \cdot 13} = \sqrt{117}$: $\qquad$ $\sin2\phi = 2\left(\dfrac{6}{\sqrt{117}}\right)\left(\dfrac{9}{\sqrt{117}}\right)$

    Next group to multiply: $\qquad$ $\sin2\phi = \dfrac{2 \cdot 6 \cdot 9}{\sqrt{117} \cdot \sqrt{117}}$

    Finally, multiply: $\qquad$ $\sin2\phi = \dfrac{2 \cdot 6 \cdot \cancel{9}}{\cancel{117}_{13}} = \dfrac{12}{13}$

11. **Answer:** $\dfrac{24}{25}$ $\quad$ If $\sin P = \dfrac{8}{10}$, then $\cos P = \dfrac{6}{10}$. (Remember the 3-4-5 right triangle? 6-8-10 is the same thing. Otherwise, use the Pythagorean theorem to find the adjacent side.)

    Now start with the identity: $\qquad$ $\sin2P = 2\sin P\cos P$

    Substitute what you know: $\qquad$ $\sin2P = 2\left(\dfrac{8}{10}\right)\left(\dfrac{6}{10}\right)$

    Next group to multiply and cancel: $\qquad$ $\sin2P = \dfrac{2 \cdot 8 \cdot 6}{10 \cdot 10} = \dfrac{\cancel{2}^1 \cdot 8 \cdot \cancel{6}^3}{\cancel{10}_5 \cdot \cancel{10}_5}$

    Finally, multiply: $\qquad$ $\sin2P = \dfrac{1 \cdot 8 \cdot 3}{5 \cdot 5} = \dfrac{24}{25}$

    Remember that sine is positive in the second quadrant, so nothing further need be done.

12. **Answer:** $-\dfrac{119}{169}$

For openers, you might use either the Pythagorean theorem or your knowledge of Pythagorean triples to find the third side of the triangle, 12. Therefore, $\cos A = \dfrac{5}{13}$.

Of course, you don't need that information if you use the identity: $\cos 2\theta = 1 - 2\sin^2\theta$

Substituting: $\cos 2\theta = 1 - 2\left(\dfrac{12}{13}\right)^2$

Clear the parentheses: $\cos 2\theta = 1 - \dfrac{2}{1} \cdot \dfrac{144}{169}$

Multiply: $\cos 2\theta = 1 - \dfrac{288}{169}$

Subtract: $\cos 2\theta = \dfrac{169}{169} - \dfrac{288}{169} = -\dfrac{119}{169}$

Since cosine is negative in quadrant II, that worked out for the best.

13. **Answer:** $\dfrac{\sqrt{14}}{7}$

First write the formula for the half angle sine identity: $\sin\dfrac{\phi}{2} = \pm\sqrt{\dfrac{1 - \cos\phi}{2}}$

Since ϕ is a positive acute angle, sine is positive: $\sin\dfrac{\phi}{2} = \sqrt{\dfrac{1 - \cos\phi}{2}}$

Next, substitute: $\sin\dfrac{\phi}{2} = \sqrt{\dfrac{1 - \dfrac{3}{7}}{2}}$

Subtract: $\sin\dfrac{\phi}{2} = \sqrt{\dfrac{\dfrac{7}{7} - \dfrac{3}{7}}{2}} = \sqrt{\dfrac{\dfrac{4}{7}}{2}}$

Divide by 2: $\sin\dfrac{\phi}{2} = \sqrt{\dfrac{4}{7} \div \dfrac{2}{1}} = \sqrt{\dfrac{4}{7} \times \dfrac{1}{2}} = \sqrt{\dfrac{4}{14}}$

Work out what's under the radical sign: $\sin\dfrac{\phi}{2} = \dfrac{\sqrt{4}}{\sqrt{14}} = \dfrac{2}{\sqrt{14}}$

Rationalize the denominator: $\sin\dfrac{\phi}{2} = \dfrac{2}{\sqrt{14}} \cdot \dfrac{\sqrt{14}}{\sqrt{14}} = \dfrac{\sqrt{14}}{7}$

14. **Answer:** $\dfrac{17}{25}$

The only identity you have that relates the two functions is: $\sin\dfrac{\phi}{2} = \pm\sqrt{\dfrac{1 - \cos\phi}{2}}$

Substitute and make the radical positive, since the angle is acute: $\dfrac{2}{5} = \sqrt{\dfrac{1 - \cos\phi}{2}}$

Square both sides of the equation: $\left(\dfrac{2}{5} = \sqrt{\dfrac{1 - \cos\phi}{2}}\right)^2 \to\to \dfrac{4}{25} = \dfrac{1 - \cos\phi}{2}$

Next, cross multiply: $25 - 25\cos\phi = 8$

Subtract 25 from both sides: $\qquad -25\cos\phi = 8 - 25 = -17$

Divide both sides by -25 and then simplify: $\quad \cos\phi = \dfrac{-17}{-25} = \dfrac{17}{25}$

15. **Answer:** $-\dfrac{\sqrt{5}}{4}$

If θ is in quadrant III, then θ is between 270° and 360°. That means $\frac{1}{2}\theta$ is greater than 135° and less than 180°. That's in quadrant II. In quadrant II, cosine is also negative.

Now write the relevant equation: $\qquad\qquad\qquad \cos\dfrac{\theta}{2} = \pm\sqrt{\dfrac{1+\cos\theta}{2}}$

Substitute and adjust the sign: $\qquad\qquad\qquad \cos\dfrac{\theta}{2} = -\sqrt{\dfrac{1+\left(-\dfrac{3}{8}\right)}{2}}$

Express both numbers to be added in like fractions: $\quad \cos\dfrac{\theta}{2} = -\sqrt{\dfrac{\dfrac{8}{8}+\left(-\dfrac{3}{8}\right)}{2}}$

Combine fractions: $\qquad\qquad\qquad\qquad \cos\dfrac{\theta}{2} = -\sqrt{\dfrac{\dfrac{5}{8}}{2}}$

Divide: $\qquad\qquad\qquad \cos\dfrac{\theta}{2} = -\sqrt{\dfrac{5}{8} \div \dfrac{2}{1}} = -\sqrt{\dfrac{5}{8} \times \dfrac{1}{2}} = -\sqrt{\dfrac{5}{16}}$

Work out what's under the radical sign: $\quad \cos\dfrac{\theta}{2} = -\dfrac{\sqrt{5}}{\sqrt{16}} = -\dfrac{\sqrt{5}}{4}$

16. **Answer:** -13

Write the formula: $\quad \tan(\alpha+\beta) = \dfrac{\tan\alpha + \tan\beta}{1 - \tan\alpha\tan\beta}$

Substitute: $\quad \tan(\alpha+\beta) = \dfrac{4+\dfrac{1}{3}}{1-\left(4\times\dfrac{1}{3}\right)}$

Combine terms: $\quad \tan(\alpha+\beta) = \dfrac{\dfrac{12}{3}+\dfrac{1}{3}}{\dfrac{3}{3}-\left(\dfrac{4}{3}\right)} = \dfrac{\dfrac{13}{3}}{-\dfrac{1}{3}}$

Divide: $\quad \tan(\alpha+\beta) = \dfrac{13}{3} \div -\dfrac{1}{3} = \dfrac{13}{\cancel{3}} \times -\dfrac{\cancel{3}}{1} = -\dfrac{13}{1} = -13$

17. **Answer:** $\dfrac{7}{6}$

First state the formula: $\quad \tan(\alpha-\beta) = \dfrac{\tan\alpha - \tan\beta}{1 + \tan\alpha\tan\beta}$

Substitute: $\quad \tan(\alpha-\beta) = \dfrac{2-\dfrac{1}{4}}{1+\left(2\times\dfrac{1}{4}\right)}$

Combine terms: $\tan(\alpha-\beta)=\dfrac{\dfrac{8}{4}-\dfrac{1}{4}}{1+\left(\dfrac{\overset{2}{\cancel{8}}}{4}\times\dfrac{1}{\cancel{4}_1}\right)}=\dfrac{\dfrac{7}{4}}{1+\dfrac{1}{2}}=\dfrac{\dfrac{7}{4}}{\dfrac{3}{2}}$

And divide: $\tan(\alpha-\beta)=\dfrac{7}{4}\div\dfrac{3}{2}=\dfrac{7}{\cancel{4}_2}\times\dfrac{\overset{1}{\cancel{2}}}{3}=\dfrac{7}{6}$

18. ***Answer:*** $-\dfrac{24}{7}$ or -3.428

    $\theta=\arcsin\dfrac{4}{5}$ means that θ is the main value of the angle whose $\sin=\dfrac{4}{5}$. From the Pythagoreqan triple, you know that the adjacent side to $\angle\theta$ must be 3, so $\tan\theta=\dfrac{4}{3}$.

    Write the formula: $\qquad\qquad\qquad\qquad\qquad\quad \tan2\theta=\dfrac{2\tan\theta}{1-\tan^2\theta}$

    Substitute: $\qquad\qquad\qquad\qquad\qquad\qquad\quad \tan2\theta=\dfrac{2\left(\dfrac{4}{3}\right)}{1-\left(\dfrac{4}{3}\right)^2}$

    Simplify numerator and denominator: $\quad \tan2\theta=\dfrac{\dfrac{8}{3}}{\dfrac{9}{9}-\dfrac{16}{9}}=\dfrac{\dfrac{8}{3}}{-\dfrac{7}{9}}$

    Finally, divide: $\qquad\qquad\qquad\qquad\quad \tan2\theta=\dfrac{8}{3}\div\left(-\dfrac{7}{9}\right)=\dfrac{8}{3}\times\left(-\dfrac{9}{7}\right)=-\dfrac{72}{21}=-\dfrac{24}{7}$

19. ***Answer:*** $\sqrt{7}$

    Select the half angle tangent formula that applies: $\quad \tan\dfrac{\theta}{2}=\pm\sqrt{\dfrac{1-\cos\theta}{1+\cos\theta}}$

    Realize that the cosine was negative because it was a quadrant III angle. Tangent in that quadrant is positive.

    Substitute and make positive: $\qquad\qquad \tan\dfrac{\theta}{2}=\sqrt{\dfrac{1-\left(-\dfrac{3}{4}\right)}{1+\left(-\dfrac{3}{4}\right)}}$

    Simplify numerator and denominator: $\quad \tan\dfrac{\theta}{2}=\sqrt{\dfrac{\dfrac{4}{4}-\left(-\dfrac{3}{4}\right)}{\dfrac{4}{4}+\left(-\dfrac{3}{4}\right)}}=\sqrt{\dfrac{\dfrac{7}{4}}{\dfrac{1}{4}}}$

    Finally, divide: $\qquad\qquad\qquad\qquad\quad \tan\dfrac{\theta}{2}=\sqrt{\dfrac{7}{4}\div\dfrac{1}{4}}=\sqrt{\dfrac{7}{\cancel{4}_1}\times\dfrac{\cancel{4}^1}{1}}=\sqrt{7}$

20. ***Answer:*** $\dfrac{\cos9w}{2}-\dfrac{\cos w}{2}$

    Pick the appropriate formula: $\quad \cos\alpha\cos\beta=\dfrac{1}{2}\left[\cos(\alpha+\beta)+\cos(\alpha-\beta)\right]$

    Substitute: $\qquad\qquad\qquad\quad \cos4w\,\cos5w=\dfrac{1}{2}\left[\cos(4w+5w)+\cos(4w-5w)\right]$

Clear parentheses: $\cos 4w \cos 5w = \frac{1}{2}\big[\cos 9w + \cos(-w)\big]$

Clear the brackets: $\dfrac{\cos 9w}{2} - \dfrac{\cos w}{2}$

21. **Answer:** sin4*d* + sin2*d*

Pick the appropriate formula: $\sin\alpha\cos\beta = \frac{1}{2}\big[\sin(\alpha+\beta) + \sin(\alpha-\beta)\big]$

Substitute: $\sin 6d\cos 2d = \frac{1}{2}\big[\sin(6d+2d) + \sin(6d-2d)\big]$

Combine terms in parentheses: $\sin 6d\cos 2d = \frac{1}{2}\big[\sin 8d + \sin 4d\big]$

Clear brackets: $\sin 6d\cos 2d = \dfrac{\sin 8d}{2} + \dfrac{\sin 4d}{2}$

Simplify: $\sin 6d\cos 2d = \sin 4d + \sin 2d$

22. **Answer:** $\dfrac{\sin 9a}{2} - \dfrac{\sin a}{2}$ or $\sin 4.5a - \dfrac{\sin a}{2}$

Pick the appropriate formula: $\cos\alpha\sin\beta = \frac{1}{2}\big[\sin(\alpha+\beta) - \sin(\alpha-\beta)\big]$

Substitute: $\cos 5a\sin 4a = \frac{1}{2}\big[\sin(5a+4a) - \sin(5a-4a)\big]$

Combine terms in parentheses: $\cos 5a\sin 4a = \frac{1}{2}\big[\sin 9a - \sin a\big]$

Clear the brackets: $\cos 5a\sin 4a = \dfrac{\sin 9a}{2} - \dfrac{\sin a}{2}$

Simplify if you like: $\cos 5a\sin 4a = \sin 4.5a - \dfrac{\sin a}{2}$

23. **Answer:** $-2\sin 55°\sin 5°$

Select the appropriate formula: $\cos\alpha - \cos\beta = -2\sin\dfrac{\alpha+\beta}{2}\sin\dfrac{\alpha-\beta}{2}$

Substitute: $\cos 60° - \cos 50° = -2\sin\dfrac{60°+50°}{2}\sin\dfrac{60°-50°}{2}$

Combine terms: $\cos 60° - \cos 50° = -2\sin\dfrac{110°}{2}\sin\dfrac{10°}{2}$

Divide: $\cos 60° - \cos 50° = -2\sin 55°\sin 5°$

24. **Answer:** 2cos6.5*u* cos1.5*u*

Select the appropriate formula: $\cos\alpha + \cos\beta = 2\cos\dfrac{\alpha+\beta}{2}\cos\dfrac{\alpha-\beta}{2}$

Substitute: $\cos 8u + \cos 5u = 2\cos\dfrac{8u+5u}{2}\cos\dfrac{8u-5u}{2}$

Combine terms: $\cos 8u + \cos 5u = 2\cos\dfrac{13u}{2}\cos\dfrac{3u}{2}$

Divide: $\cos 8u + \cos 5u = 2\cos 6.5u\cos 1.5u$

25. **Answer:** 2cos50°sin5°

Select the appropriate formula: $\sin\alpha - \sin\beta = 2\cos\dfrac{\alpha+\beta}{2}\sin\dfrac{\alpha-\beta}{2}$

Substitute: $\sin55° - \sin45° = 2\cos\dfrac{55°+45°}{2}\sin\dfrac{55°-45°}{2}$

Combine terms: $\sin55° - \sin45° = 2\cos\dfrac{100°}{2}\sin\dfrac{10°}{2}$

Divide: $\sin55° - 45° = 2\cos50°\sin5°$

Supplemental Chapter Problems

Solve these problems for even more practice applying the skills from this chapter. The Answer section will direct you to where you need to review.

Problems

1.   Write the reciprocal identity for secant.

2.   Write the ratio identity for tangent.

3.   Write the cofunction identity for sine in both ratio and complementary form.

4.   Write an expression that is always equivalent to $\cos(-\theta)$.

5.   What is the value of $13\sin^2\lambda + 13\cos^2\lambda$?

6.   Rewrite the expression, $\tan\phi(\cos\phi - \csc\phi)$, in simplest form.

7.   Evaluate the expression $9\csc^2\theta - 9\cot^2\theta$.

8.   Change $\cos100°\cos60° + \sin100°\sin60°$ into a trigonometric function in a single variable and evaluate it.

9.   Change $\sin35°\cos85° + \cos35°\sin85°$ into a trigonometric function in a single variable and evaluate it.

10.   If $\cos\phi = \dfrac{6}{\sqrt{61}}$, find the value of $\sin2\phi$.

11.   If A is a second quadrant angle, and $\sin A = \dfrac{9}{15}$, what is $\cos 2A$?

12.   If P is in the second quadrant and $\sin P = \dfrac{5}{8}$, find $\sin2P$.

13.   If $\cos\theta = -\dfrac{3}{8}$ and θ is in quadrant III, find $\cos\dfrac{1}{2}\theta$.

14.   If ϕ is a positive acute angle and $\cos\phi = \dfrac{2}{9}$, find $\sin\dfrac{\phi}{2}$.

15.   If $\sin\dfrac{1}{2}\theta = \dfrac{1}{6}$ and θ is an acute angle, find $\cos\theta$.

16. If $\tan\alpha = 6$ and $\tan\beta = \frac{1}{2}$, what is the tangent of their sum?

17. If $\tan\alpha = 4$ and $\tan\beta = \frac{1}{4}$, what is the tangent of their difference $(\alpha - \beta)$?

18. If $\theta = \arcsin\frac{5}{13}$, find tan 2.

19. If the cosine of a second quadrant angle is $\frac{3}{7}$, find the tangent of an angle half that size.

20. Express cos6m sin4m as a sum.

21. Express sin9b sin7b as a sum.

22. Express sin6x cos4x as a sum.

23. Write the sum sin3n + sin7n as a product.

24. Write the difference sin65° − sin35° as a product.

25. Write the sum sin6w + sin8w as a product.

Answers

1. $\sec\theta = \frac{1}{\cos\theta}$ (Reciprocal Identities, p. 137)

2. $\tan\theta = \frac{\sin\theta}{\cos\theta}$ (Ratio Identities, p. 138)

3. $\sin A = \frac{a}{c} = \cos B$, $\sin\theta = \cos(90° - \theta)$ (Cofunction Identities, p. 139)

4. $-\cos\theta$ (Identities for Negatives, p. 139)

5. 13   (Pythagorean Identities, p. 140)

6. $\sin\phi - \sec\phi$ (Pythagorean Identities, p. 140)

7. 9   (Pythagorean Identities, p. 140)

8. 0.766   (Addition and Subtraction Identities, p. 143)

9. 0.866   (Addition and Subtraction Identities, p. 143)

10. $\frac{60}{61}$ (Double Angle Identities, p. 144)

11. $-\frac{7}{25}$ (Double Angle Identities, p. 144)

12. $\frac{5\sqrt{39}}{32}$ (Double Angle Identities, p. 144)

13. $\frac{-\sqrt{5}}{4}$ (Half Angle Identities, p. 146)

14. $\dfrac{\sqrt{14}}{3}$ (Half Angle Identities, p. 146)

15. $\dfrac{17}{18}$ (Half Angle Identities, p. 146)

16. $-\dfrac{13}{4}$ or -3.25 (Tangent Identities, p. 150)

17. $\dfrac{15}{8}$ or 1.875 (Tangent Identities, p. 150)

18. $\dfrac{120}{119}$ or 1.008 (Tangent Identities, p. 150)

19. $\dfrac{\sqrt{10}}{5}$ (Tangent Identities, p. 150)

20. $\sin 5m - \sin m$ (Product-Sum Identities, p. 152)

21. $\cos b - \cos 8b$ (Product-Sum Identities, p. 152)

22. $\sin 5x + \sin x$ (Product-Sum Identities, p. 152)

23. $2\cos 5n \cos(-4n)$ (Sum-Product Identities, p. 153)

24. $2\cos 50° \sin 15°$ (Sum-Product Identities, p. 153)

25. $2\sin 7w \cos(-w)$ (Sum-Product Identities, p. 153)

Chapter 5

Vectors

Vectors versus Scalars

In real life, many measures such as length, mass, age, time, volume, and value have magnitude (a fancy word for size) alone. Such measures are known as **scalars,** a word that you could easily go through life without ever encountering. Other measures, such as force and velocity have both magnitude and direction. Measures with both force and direction are known as **vectors,** and they are typically represented by arrows. One common use of vectors is finding the actual direction and speed at which one would have to row a boat in order to get across a river to a certain point when the river is flowing downstream at a certain velocity. For example, in the following figure, to get from A to B, you would row along $\overrightarrow{AC}$.

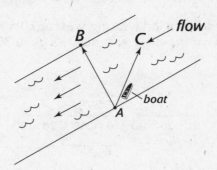

To get from A to B, row along $\overrightarrow{AC}$.

Notice that the vector $\overrightarrow{AC}$ is written the same way as you would represent a ray in geometry. You need to know whether you are dealing with rays or vectors. In vector notation, the **directed line segment** $\overrightarrow{AC}$ has an initial point A and a terminal point, C.

Although you are no doubt accustomed to seeing speed limit signs on highways, speed is a scalar, not a vector. Velocity, a vector, has both speed and direction. From now on, we'll use a boldface letter to represent vectors, such as **v** and **u.** Two vectors have the same direction if they are parallel and point the same way. The magnitude of a vector is indicated by the length of the directed line segment.

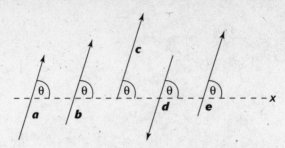

Four equivalent vectors and one opposite vector.

In the preceding figure, vectors **a, b, c,** and **e** are **equivalent vectors.** That is they are all of the same magnitude and direction. Vector **d,** however, is an **opposite vector** of the same magnitude, parallel to the four others, but pointing in the opposite direction.

The following problems make reference to directions with respect to the points on a compass. Conventionally, the points of the compass show north as up, south as down, west as left, and east as right.

Example Problems

These problems show the answers and solutions.

1. Draw a vector **w** to represent a wind blowing west at 20 miles per hour for one hour, and a vector **c** representing a car travelling east that has gone 30 miles.

    answer:

    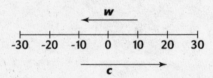

    Note that since a starting point was not specified for either vector, it may begin anywhere as long as it moves in the specified direction and is of the specified magnitude.

2. Which of the following may be represented by vectors?
    a. your height
    b. your weight
    c. the velocity of a meteor
    d. the time school begins
    e. a line segment on a map connecting New York City and Chicago
    f. your pulse
    g. the wind
    h. a thrown football

answer: b, c, g, h

a. A tape measure can find your height whether you're standing or lying down. It has no direction.

b. Try putting a scale on the wall and weighing yourself. Weight is a function of gravity with a magnitude and a direction—down.

c. Velocity has already been mentioned as having speed and direction.

d.–f. d, e, and f all have magnitude but no direction. The line segment was not pointing to either end.

g.–h. g and h both have magnitude and direction.

Vector Addition Triangle/The Tip-Tail Rule

Vector addition may be shown in two different ways. The following figure shows a triangle of vectors representing two forces acting on a body at O. $\overrightarrow{OB}$ represents a force pulling up and slightly to the right, **v.** $\overrightarrow{OA}$ is a force pulling on the same body due right, **u.**

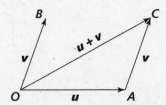

The tip-tail rule illustrated.

The **tip–tail rule** says: To add two vectors, place the tip of one to the tail of an equivalent of the other, as is represented by $\overrightarrow{AC}$. Note that $\overrightarrow{AC}$ is the same size and goes in the same direction as $\overrightarrow{OB}$. It is equivalent to $\overrightarrow{OB}$ but has been moved to place its tail at the tip of $\overrightarrow{OA}$. The result of this vector addition is $\overrightarrow{OC}$, known as **the resultant,** which connects the initial point of $\overrightarrow{OA}$ to the terminal point of $\overrightarrow{AC}$. $\overrightarrow{OA}$ and $\overrightarrow{OB}$ are referred to as **components.**

To better understand vector addition, consider a city block with a single large building on it. You want to travel from O to C, but the building is in the way, so you must go around it by way of A. Your trip from O to C via A results in your ending up where you would have been if the building hadn't been in the way. Also, remember that the rule that says two sides of a triangle must add up to a sum greater than the third side is still in effect. When adding two vectors, you are not adding the lengths of the two sides, but rather two directional sides. The directions play a part in the vector addition and in the direction of the resultant.

Example Problems

These problems show the answers and solutions.

1. Draw the resultant of vectors $\overrightarrow{AB}$ and $\overrightarrow{CD}$ as shown in the following figure.

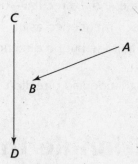

Vectors for problem 1.

answer:

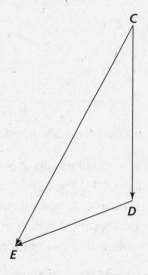

One solution to problem 1.

First draw $\overrightarrow{CD}$. Then, at the tip of $\overrightarrow{CD}$ place $\overrightarrow{DE}$, the equivalent vector of $\overrightarrow{AB}$. Finally, connect the initial point of $\overrightarrow{CD}$ to the terminal point of $\overrightarrow{DE}$ to get resultant $\overrightarrow{CE}$.

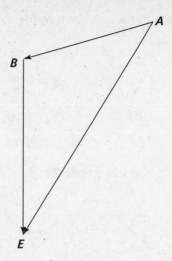

Another solution to problem 1.

By the way, you could also have started with $\overrightarrow{AB}$ and then placed $\overrightarrow{BE}$ (the equivalent vector of $\overrightarrow{CD}$), as in the previous figure. The resultant would have been $\overrightarrow{AE}$.

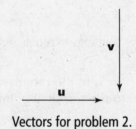

Vectors for problem 2.

2. Draw the resultant of vectors **v** and **u** as shown in the previous figure.

answer:

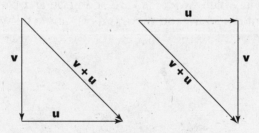

Two solutions to problem 2.

Bring the tip of vector **v** to the tail of **u** or the other way around. In either case, connect the initial point of the first vector to the terminal point of the second to draw the resultant.

3. Draw the resultant of vectors **v** and **u** as shown in the following figure.

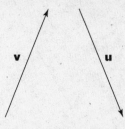

Vectors for problem 3.

answer:

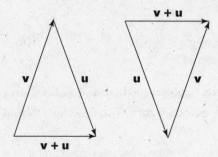

Two solutions to problem 3.

Bring the tip of vector **v** to the tail of **u** or the other way around. In either case, connect the initial point of the first vector to the terminal point of the second to draw the resultant.

Parallelogram of Forces

Much more popular with physicists who deal with analyzing forces everyday than the vector triangle is the **vector parallelogram,** or as it is more popularly known, the **parallelogram of forces.** Consider a situation in which two forces are acting on an object, *O*; one component is 25 lbs.; the other component is 15 lbs.; and there is a 60° angle between the two forces. See the following figure.

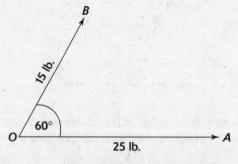

Two forces acting at a 60° angle.

By drawing $\overline{AC} \parallel \overline{OB}$ and $\overline{BC} \parallel \overline{OA}$, you form a parallelogram. The diagonal of that parallelogram, $\overline{OC}$, is the resultant of the two components, as in the following figure.

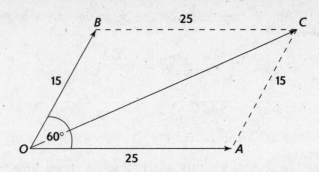

The diagonal from the initial points of the vectors is the resultant.

Remember the Law of Cosines?

$$a^2 = b^2 + c^2 - 2bc\cos A \text{ or } \cos A = \frac{b^2 + c^2 - a^2}{2bc}$$

You can use it to determine the magnitude of resultant *OC*.

Example Problems
These problems show the answers and solutions.

1. Solve the parallelogram of forces in the preceding figure.

   ***answer:* 35 lbs. at ≈ 21.8°**

   Remember that the Law of Cosines can be used with any general triangle. The 60° angle between the vectors is not usable, since it is not contained wholly within either triangle. However, consecutive angles of a parallelogram are supplementary, so that means $m\angle A = 120°$. $180° - 120°$ gives you a reference angle of 60°, and cosine is negative in quadrant II.

   State the formula: $\cos A = \dfrac{b^2 + c^2 - a^2}{2bc}$

   Substitute: $-\dfrac{1}{2} = \dfrac{15^2 + 25^2 - a^2}{15 \cdot 25 \cdot 2}$

   Multiply: $-\dfrac{1}{2} = \dfrac{225 + 625 - a^2}{30 \cdot 25}$

   Combine like terms: $-\dfrac{1}{2} = \dfrac{850 - a^2}{750}$

   Cross multiply: $2(850 - a^2) = -1 \cdot 750$

   Divde both sides by 2: $850 - a^2 = -375$

   Collect terms: $-a^2 = -375 - 850$

   Add; multiply by −1: $a^2 = 1225$

   $\sqrt{\text{both sides}}$: $a = 35 \text{ lbs.}$

So the resultant is 35 lbs. in magnitude. Still to be settled is the angle $\angle COA$. To find it, use the Law of Sines.

State the formula: $\qquad \dfrac{a}{\sin A} = \dfrac{\text{side opposite } \angle COA}{\sin \angle COA}$

Substitute: $\qquad\qquad \dfrac{35}{\sin 120°} = \dfrac{15}{\sin \angle COA}$

Cross multiply: $35 \sin \angle COA = 15 \sin 120°$

The quadrant II reference angle for 120° is 60°, and sine is positive in quadrant II, so,

$35 \sin \angle COA = 15 \sin 60°$.

Substitute and divide both sides by 35: $\sin \angle COA = \dfrac{15 \cdot 0.867}{35} = \dfrac{13.005}{35}$

Divide by 35: $\qquad\qquad\qquad\qquad\qquad\quad \sin \angle COA = 0372$

Find arcsin or $\sin^{-1} 0.372$: $\qquad\qquad\qquad\qquad \angle COA \approx 21.8°$

2.   A force of 12 kg and a force of 16 kg are acting on a body at an angle of 90°. What are the magnitude and angle of the resultant with respect to the 16 kg force?

    answer: 20 kg, $\approx$ 36.9°

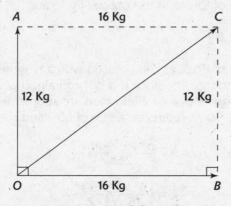

Forces acting at right angles.

Always start by drawing a diagram. It just so happens you have one in the preceding figure. $\overrightarrow{OB}$ represents the 16 kg force, and $\overrightarrow{OA}$ represents the 12 kg force. O is the object upon which the forces are acting, and $\overrightarrow{OC}$ is the resultant force. Notice the corners of this parallelogram are right angles, so the parallelogram of forces is, in fact, a rectangle of forces. That means you can find $\overrightarrow{OC}$ through the magic of the Pythagorean theorem. This is by way of the old maxim: Never use an elephant gun when a pea-shooter will do the trick.*

State the equation: $\qquad\qquad\qquad c^2 = a^2 + b^2$

Substitute: $\qquad\qquad\qquad\qquad c^2 = 12^2 + 16^2$

Square the appropriate numbers: $c^2 = 144 + 256$

---

*For those who don't know what a pea shooter is, it's a plastic straw used to blow dried peas at fellow students when the teacher isn't looking—not that you or I would ever do such a thing.

Add: $c^2 = 400$

Take the square root of both sides: $c = 20$

So the resultant is 20 kg in magnitude. To find its angle with respect to the 16 kg force, you must find $\angle COB$.

You know that the tangent of $\angle COB = \dfrac{12}{16} = \dfrac{3}{4} = 0.75$

Look that up in the table or on your calculator, and you'll find $\sin^{-1} 0.75 \approx 36.9°$.

3. An airplane is trying to fly due west at 500 mph. Because of a wind from the south, the plane's actual path is 15° north of west. Assuming the wind to be blowing at a steady pace, what is its magnitude, to the nearest whole mph?

    answer: **134 mph**

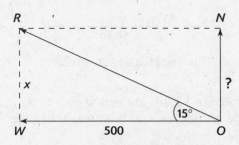

Plane is blown north of west.

First, start with the diagram, as in the preceding figure where x is the equivalent vector of the wind, $\vec{ON}$. With respect to the angle at O, the relevant relationship is tangent.

State the equation: $\tan\theta = \dfrac{\text{opposite}}{\text{adjacent}}$

Substitute: $\tan 15° = \dfrac{x}{500}$

Cross multiply: $x = 500\tan 15°$

Find $\tan 15°$: $x = 500(0.268)$

Multiply: $x = 134$ mph.

Work Problems

Use these problems to give yourself additional practice.

1. Use the tip-tail rule to draw the two solutions for the resultant for the vectors in the following figure.

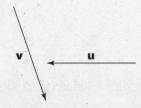

2. Use the tip-tail rule to draw the two solutions for the resultant for the vectors in the following figure.

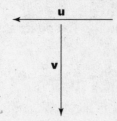

3. Solve the following figure using the parallelogram of forces.

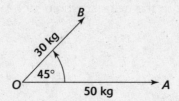

Two forces acting on an object.

4. A force of 9 lbs. and a force of 12 lbs. are acting on a body at an angle of 90°. What are the magnitude and angle of the resultant with respect to the 12-pound force?

Worked Solutions

1. The two possible solutions are shown in the following figure.

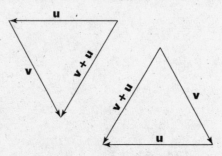

Two possible solutions for problem 1.

2. The two possible solutions are shown in the following figure.

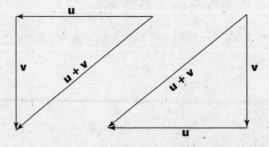

Two possible solutions for problem 2.

3. **About 66.8 kg at ≈ 18.5°**   First, you need to complete the diagram, as in the following figure.

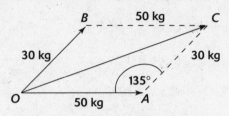

Completing the parallelogram of forces.

Remember that the Law of Cosines can be used with any general triangle. The 45° angle between the vectors is not usable, since it is not contained wholly within either triangle. However, consecutive angles of a parallelogram are supplementary, so that means $m\angle A = 135°$. $180° - 135°$ gives us a reference angle of 45°, and cosine is negative in quadrant II.

If you remember that $\cos 45° = \dfrac{\sqrt{2}}{2}$, use it; otherwise, 0.707 can be found with your calculator or trig tables.

State the formula:

$$\cos A = \frac{b^2 + c^2 - a^2}{2bc}$$

Substitute:

$$-\frac{\sqrt{2}}{2} = \frac{30^2 + 50^2 - a^2}{30 \cdot 50 \cdot 2}$$

Multiply:

$$-\frac{\sqrt{2}}{2} = \frac{900 + 2500 - a^2}{30 \cdot 100}$$

Combine like terms:

$$-\frac{\sqrt{2}}{2} = \frac{3400 - a^2}{3000}$$

Cross multiply: $-2(3400 - a^2) = 3000\sqrt{2}$

Divide both sides by −2: $3400 - a^2 = -1500\sqrt{2}$

The $\sqrt{2}$ has to go: $3400 - a^2 = -1500 \cdot 0.707 = -1060.5$

Collect terms: $-a^2 = -1060.5 - 3400$

Add; multiply by −1: $a^2 = 1225 \cdot a^2 = 4460.5$

$\sqrt{\ }$ both sides: $a \approx 66.8$ kg.

So the resultant is about 67 kg in magnitude. Still to be settled is the angle $\angle COA$. To find it, use the Law of Sines.

State the formula: $\dfrac{a}{\sin A} = \dfrac{\text{side opposite } \angle COA}{\sin \angle COA}$

Substitute: $\dfrac{67}{\sin 135°} = \dfrac{30}{\sin \angle COA}$

Cross multiply: $67 \sin \angle COA = 30 \sin 135°$

The quadrant II reference angle for 135° is 45°, and sine is positive in quadrant II, so,

Substitute: $67\sin\angle COA = 30\sin45° = 30 \cdot 0.707$

Divide both sides by 67: $\sin\angle COA = \dfrac{21.21}{67}$

Divide: $\sin\angle COA = 0.317$

Find arcsin or $\sin^{-1}$ 0.317: $COA \approx 18.5°$

4. **15 lbs., ≈ 37°**

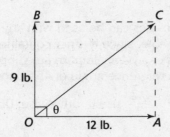

Always start by drawing a diagram like the preceding . $\overrightarrow{OB}$ represents the 9 lb. force, and $\overrightarrow{OA}$ represents the 12 lb. force. O is the object upon which the forces are acting, and $\overrightarrow{OC}$ is the resultant force. Notice the corners of this parallelogram are right angles, so the parallelogram of forces is, in fact, a rectangle of forces. That means you can find $\overrightarrow{OC}$ through the magic of the Pythagorean theorem.

State the equation: $c^2 = a^2 + b^2$

Substitute: $c^2 = 9^2 + 12^2$

Square the appropriate numbers: $c^2 = 81 + 144$

Add: $c^2 = 225$

Take the square root of both sides: $c = 15$

So the resultant is 15 lbs. in magnitude. To find its angle with respect to the 12 lb. force, you must find $\angle COB$.

You know that the tangent of $\angle COB = \dfrac{9}{12} = \dfrac{3}{4} = 0.75$. Look that up in the table or on your calculator, and you'll find $\tan^{-1} 0.75 \approx 37°$.

Vectors in the Rectangular Coordinate System

If a vector is drawn so that its initial point is at the origin of the Cartesian coordinate plane, it is said to be in standard position.

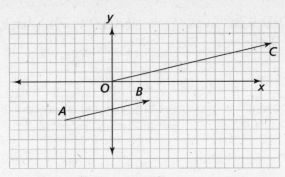

Equivalent vectors with $\overrightarrow{OC}$ in standard position.

In the preceding figure, vector $\overrightarrow{AB}$ is equivalent to vector $\overrightarrow{OC}$. Since $\overrightarrow{OC}$ has its initial point at the origin, it is said to be the standard vector for $\overrightarrow{AB}$. **Position vector, radius vector,** and **centered vector** are three other names sometimes used to refer to the standard vector. Any vector in the coordinate plane that does not have its starting point at the origin is known as a **free vector.**

Vector $\overrightarrow{OC}$ is the standard vector for all vectors in the plane that have the same direction and magnitude. In order to identify a standard vector in the coordinate plane, only the coordinates of the terminal point need to be named, since the original point is at the origin. If the coordinates of point A are (x_a, y_a) and the coordinates of point B are (x_b, y_b), then the coordinates of point C are $(x_b - x_a, y_b - y_a)$. In the case of the preceding figure, that would be $(4 - -5, -2 - -4)$, which computes to $(9, 2)$.

Example Problems

These problems show the answers and solutions.

1. If the endpoints of vector $\overrightarrow{AB}$ are A $(-3, -5)$ and B $(4, 7)$, what are the coordinates of standard vector $\overrightarrow{OP}$ if $\overrightarrow{OP} = \overrightarrow{AB}$?

    ***answer:* (7, 12)**

    Remember, if the coordinates of point A are (x_a, y_a) and the coordinates of point B are (x_b, y_b), then the coordinates of point P are $(x_b - x_a, y_b - y_a)$.

    Substitute: $(4 - -3, 7 - -5)$

    Add: $(7, 12)$

    That, of course, is the terminal point of $\overrightarrow{OP}$, its initial point being at the origin.

2. If the endpoints of vector $\overrightarrow{CD}$ are C $(-8, -6)$ and D $(-1, 3)$, what are the coordinates of standard vector $\overrightarrow{OM}$ if $\overrightarrow{OM} = \overrightarrow{CD}$?

    ***answer:* (7, 9)**

    Since the coordinates of point C are (x_a, y_a) and the coordinates of point D are (x_b, y_b), then the coordinates of point M are $(x_b - x_a, y_b - y_a)$.

    Substitute: $(-1 - -8, 3 - -6)$

    Add: $(7, 9)$

3. If the endpoints of vector $\overrightarrow{AB}$ are A (3, −5) and B (11, −9), what are the coordinates of standard vector $\overrightarrow{OP}$ if $\overrightarrow{OP} = \overrightarrow{AB}$?

    answer: (8, −4)

    Remember, if the coordinates of point A are (x_a, y_a) and the coordinates of point B are (x_b, y_b), then the coordinates of point P are $(x_b − x_a, y_b − y_a)$.

    Substitute: $(11 − 3, −9 − −5)$

    Add: $(8, −4)$

Resolution of Vectors

Resolution of vectors is better known in the science of physics as **resolution of forces.** The typical problem presents an oblique vector in the Cartesian plane (one neither horizontal nor vertical) and resolves it into its rectangular, that is to say its horizontal and vertical components. It is common for electrical engineers to denote a vector of magnitude s and an angle with the horizontal θ as $s\angle\theta$. That notation is used in the problems that follow.

Example Problems

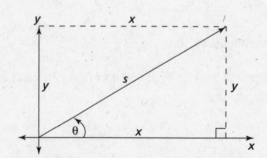

Model for resolution of vectors problems.

These problems show the answers and solutions. Refer to the preceding figure for the model.

1. Find the rectangular components of the vector $12\angle 30°$ to the nearest integer.

    answer: x = 10, y = 6

    Start with sine: $\sin 30° = \dfrac{\text{opposite}}{\text{hypotenuse}}$

    Substitute: $\dfrac{1}{2} = \dfrac{y}{12}$

    Cross multiply: $2y = 12$

    Divide both sides by 2: $y = 6$

Next, use either tangent or cosine: $\cos 30° = \dfrac{\text{adjacent}}{\text{hypotenuse}}$

Substitute, remembering that $\cos 30° \dfrac{\sqrt{3}}{2}$: $\dfrac{\sqrt{3}}{2} = \dfrac{x}{12}$

Cross multiply: $2x = 12\sqrt{3}$

Divide both sides by 2: $x = 6\sqrt{3}$ or 10

2. The rate and angle with the bank of a boat crossing a stream in miles per hour is $25\angle 40°$. What are the cross-stream and current components of the boat's velocity?

   answer: **Current is 19 mph; cross-stream rate is 16 mph.**

   Since the angle with the bank is 25°, you can safely assume from referencing the preceding figure that that the x component is the downstream one, that is, the current, and that the cross-stream rate is the y component.

   Start with sine: $\sin 40° = \dfrac{\text{opposite}}{\text{hypotenuse}}$

   Substitute: $\dfrac{0.643}{1} = \dfrac{y}{25}$

   Cross multiply: $y = 25 \cdot 0.643$

   Multiply and round: $y = 16.075 \approx 16$

   Next, use cosine, since the y-value is not exact: $\cos 40° = \dfrac{\text{adjacent}}{\text{hypotenuse}}$

   Substitute, remembering that: $\dfrac{0.766}{1} = \dfrac{x}{25}$

   Cross multiply: $x = 25 \cdot 0.766$

   Multiply and round: $x = 25 \cdot 0.766 = 19.15 \approx 19$

3. The magnitude of the horizontal component of a force is twice the magnitude of the vertical component. Find the angle the resultant force makes with the larger component.

   answer: $\approx 27°$

   Let the angle be ϕ, the vertical component be x, and the horizontal component be $2x$.

   Use the tangent function, since: $\tan\phi = \dfrac{\text{opposite}}{\text{adjacent}}$

   Substitute: $\tan\phi = \dfrac{x}{2x}$

   But the xs cancel each other, so: $\tan\phi = \dfrac{1\!\!\!/x}{2\!\!\!/x} = \dfrac{1}{2}$

   Now find the angle whose tangent is 0.5: $\tan\phi = 26.57 \approx 27° \ (26.57)$

Work Problems

Use these problems to give yourself additional practice.

1. If the endpoints of vector $\overrightarrow{AB}$ are A (−5, −8) and B (4, 7), what are the coordinates of standard vector $\overrightarrow{OP}$ if $\overrightarrow{OP} = \overrightarrow{AB}$?

2. If the endpoints of vector $\overrightarrow{CD}$ are C (−12, −8) and D (−2, 4), what are the coordinates of standard vector $\overrightarrow{OM}$ if $\overrightarrow{OM} = \overrightarrow{CD}$?

3. Find the rectangular components of the vector 16∠30° to the nearest integer.

4. The magnitude of the vertical component of a force is three times the magnitude of the horizontal component. Find the angle the resultant force makes with the smaller component.

Worked Solutions

1. **(9, 15)**   Follow the formula that says if the coordinates of point A are (x_a, y_a) and the coordinates of point B are (x_b, y_b), then the coordinates of point P are $(x_b - x_a, y_b - y_a)$.

    Substitute: $(4 - -5, 7 - -8)$

    Add: $(9, 15)$

    That, of course, is the terminal point of $\overrightarrow{OP}$; its initial point being at the origin.

2. **(10, 12)**   Since the coordinates of point C are (x_a, y_a) and the coordinates of point D are (x_b, y_b), then the coordinates of point M are $(x_b - x_a, y_b - y_a)$.

    Substitute: $(-2 - -12, 4 - -8)$

    Add: $(10, 12)$

3. **$x = 14$, $y = 8$**   Start with sine: $\sin 30° = \dfrac{\text{opposite}}{\text{hypotenuse}}$

    Substitute: $\dfrac{1}{2} = \dfrac{y}{16}$

    Cross multiply: $2y = 16$

    Divide both sides by 2: $y = 8$

    Next, use either tangent or cosine: $\cos 30° = \dfrac{\text{adjacent}}{\text{hypotenuse}}$

    Substitute remembering that $\cos 30° = \dfrac{\sqrt{3}}{2}$: $\dfrac{\sqrt{3}}{2} = \dfrac{x}{16}$

    Cross multiply: $2x = 16\sqrt{3}$

    Divide both sides by 2: $x = 8\sqrt{3}$ or 14

4. **≈ 72°**   Let the angle be ϕ, the horizontal component be x, and the vertical component be $3x$.

    Use the tangent function, since: $\tan\phi = \dfrac{\text{opposite}}{\text{adjacent}}$

    Substitute: $\tan\phi = \dfrac{3x}{x}$

    But the xs cancel each other, so: $\tan\phi = \dfrac{3\cancel{x}}{1\cancel{x}} = 3$

    Now find the angle whose tangent is 3: $\tan\phi = 71.57 \approx 72°$

Algebraic Addition of Vectors

An **algebraic vector** is an ordered pair of numbers. One corresponding to vector **a** in the following figure is represented by the ordered pair, (a, b); a and b are known as the **components** of the vector, a being the horizontal component and b being the vertical one.

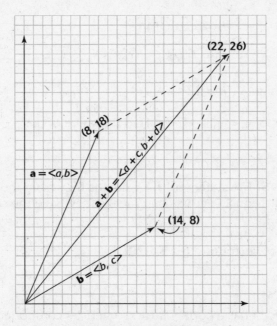

Algebraic vector addition.

Here are a few basic facts that will help when working with algebraic vectors.

❑ Two vectors with the same components are known as equal vectors.

❑ If both components of a vector are equal to zero, the vector is called the **zero vector.**

❑ The **magnitude** of a vector, **v** = (a, b) is found by doing $|v| = \sqrt{a^2 + b^2}$. This should come as no surprise, since a and b are the horizontal and vertical components of the vector. Think of the rectangular parallelogram of forces in standard position with the diagonal being the hypotenuse of the right triangle on the right. If you're having trouble visualizing that, go back and look at the figure on page 178.

❑ **Vector addition** can be algebraically defined as adding corresponding components of vectors. Look at the preceding figure. Note how the horizontal components of **a** and **b** have been added to get the horizontal component of the resultant, **a** + **b.** Then the vertical components were added to get the vertical component of **a** + **b.** Prove it to yourself by using the numerical coordinates of the two vectors and the resultant.

Example Problems

These problems show the answers and solutions.

1. Find the magnitude of vector **h** = (−5, 12)

    ***answer:* 13**

    Remember that a vector has a horizontal and a vertical component that are −5 and 12, respectively. The resultant may be represented as the hypotenuse of a right triangle.

    You can use the Pythagorean theorem or the version of the distance formula shown previously.

    Write the formula: $|h| = \sqrt{a^2 + b^2}$

    Substitute: $|h| = \sqrt{(-5)^2 + 12^2}$

    Square: $|h| = \sqrt{25 + 144}$

    Combine terms: $|h| = \sqrt{169}$

    Remove the radical sign: $|h| = 13$

2. Vector **f** = (5, 7); vector **g** = (c, d). **f** = **g**. Find the value of d.

    ***answer:* 7**

    And the value of c is 5. The statement **f** = **g** means they have the same (or identical) components. They are equal vectors. That means they have the same horizontal and vertical components.

3. Find the magnitude of vector **m** = (−15, −20).

    ***answer:* 25**

    Remember that a vector has a horizontal and a vertical component that are −15 and −20, respectively. The resultant may be represented as the hypotenuse of a right triangle.

    You can use the Pythagorean theorem or the version of the distance formula used previously.

    Write the formula: $|m| = \sqrt{a^2 + b^2}$

    Substitute: $|m| = \sqrt{(-15)^2 + (-20)^2}$

    Square: $|m| = \sqrt{225 + 400}$

    Combine terms: $|m| = \sqrt{625}$

    Remove the radical sign: $|m| = 25$

Scalar Multiplication

To multiply a vector by a scalar (an ordinary number), both components of the vector are multiplied by the scalar.

For instance, using the constant, k: $k(\mathbf{a}) = k(x_a, y_a)$.

In turn: $k(x_a, y_a) = (kx_a, ky_a)$.

Multiplication by a scalar is commutative. That means it doesn't matter in which order the factors appear.

Vector × Scalar = Scalar × Vector.

In notation: $k(\mathbf{a}) = \mathbf{a}(k) = (kx_a, ky_a)$

Another way to state this is $k(a, b) = (ak, bk) = (ka, kb)$ with the last form being preferred, since it is customary for constant to precede variable.

Example Problems

These problems show the answers and solutions.

1. Multiply $\mathbf{v}$ (−3, 7) by 8.

    answer: $8\mathbf{v} = (-24, 56)$

    That didn't require a whole lot of thought. You simply multiply both components by the scalar, 8.

2. If $\mathbf{e} = (c, d)$ and $\mathbf{f} = (g, h)$, find the product of 9 and $\mathbf{e} + \mathbf{f}$.

    answer: $9(\mathbf{e} + \mathbf{f}) = (9c, 9d) + (9g, 9h)$

    Just remember to multiply each component by 9.

3. If $\mathbf{m} = (7, -8)$ and $\mathbf{n} = (-7, 12)$, find the product of −7 and $\mathbf{m} - \mathbf{n}$.

    answer: $-7\ (\mathbf{m} - \mathbf{n}) = (-49, 56) - (49, -84)$

    Just remember to multiply each component by −7, paying careful attention to the signs.

Dot Products

The **dot product** is also known as **scalar multiplication,** not to be confused with *multiplication by a scalar.* It is named for the multiplication dot placed between two vectors when multiplying them: $\mathbf{v} \cdot \mathbf{u}$ is the dot product of $\mathbf{v}$ and $\mathbf{u}$. If the components of $\mathbf{v}$ are (x_v, y_v) and the components of $\mathbf{u}$ are (x_u, y_u), then the dot product $\mathbf{v} \cdot \mathbf{u} = x_u x_v + y_u y_v$ − the sum of the product of the horizontal components and the product of the vertical components. You may have noticed that the order of the subscripts was reversed in the product. That's because it is customary to list variables in alphabetical order. To make a long story short, since vector multiplication, like all real number multiplication, is commutative, $\mathbf{v} \cdot \mathbf{u} = \mathbf{u} \cdot \mathbf{v}$.

Example Problems

These problems show the answers and solutions.

1. Multiply **v** (−3, 7) by **u** (−6, 4).

    answer: v · u = 46

    By formula: $\mathbf{v} \cdot \mathbf{u} = x_u x_v + y_u y_v$

    Substitute: $\mathbf{v} \cdot \mathbf{u} = -3 \cdot -6 + 7 \cdot 4$

    Multiply: $\mathbf{v} \cdot \mathbf{u} = 18 + 28$

    And add: $\mathbf{v} \cdot \mathbf{u} = 46$

2. If **n** = (c, d) and **o** = (g, h), find the product **n · o**.

    answer: n · o = $cg + dh$

    By formula: $\mathbf{n} \cdot \mathbf{o} = x_n x_o + y_n y_o$

    Substitute: $\mathbf{n} \cdot \mathbf{o} = cg + dh$

    And that's as far as you can go.

3. If **m** = (7, −8) and **n** = (−7, 12), find the dot product of **m** and **n**.

    answer: m · n = −145

    By formula: $\mathbf{m} \cdot \mathbf{n} = x_m x_n + y_m y_n$

    Substitute: $\mathbf{m} \cdot \mathbf{n} = (7)(-7) + (-8)(12)$

    Multiply: $\mathbf{m} \cdot \mathbf{n} = -49 + -96$

    And add: $\mathbf{m} \cdot \mathbf{n} = -145$

Work Problems

Use these problems to give yourself additional practice.

1. Find the magnitude of vector **p** = (−10, 24).

2. Find the magnitude of vector **w** = (7, 8).

3. If **e** = (c, d) and **f** = (g, h), find the product of −8 and **e** + **f**.

4. If **r** = (6, 9) and **s** = (8, −10), find the product of −4 and **r** − **s**.

5. Multiply **q** (6, 5) by **r** (−5, 6).

6. If **m** = (4, −6) and **n** = (5, −7), find the dot product of **m** and **n**.

Worked Solutions

1. . 26 Remember that a vector has a horizontal and a vertical component, which are −10 and 24, respectively. The resultant may be represented as the hypotenuse of a right triangle.

   You can use the Pythagorean theorem or the distance formula.

   Write the formula: $|p| = \sqrt{a^2 + b^2}$

   Substitute: $|p| = \sqrt{(-10)^2 + 24^2}$

   Square: $|p| = \sqrt{100 + 576}$

   Combine terms: $|p| = \sqrt{676}$

   Remove the radical sign: $|p| = 26$

2. $\sqrt{113}$ or 10.63 Remember that a vector has a horizontal and a vertical component, which are 7 and 8, respectively. The resultant may be represented as the hypotenuse of a right triangle.

   You can use the Pythagorean theorem or the distance formula.

   Write the formula: $|w| = \sqrt{a^2 + b^2}$

   Substitute: $|w| = \sqrt{7^2 + 8^2}$

   Square: $|w| = \sqrt{49 + 64}$

   Combine terms: $|w| = \sqrt{113}$

   That's as far as it can be simplified, so leave it in radical form, or calculate $\sqrt{113} = 10.63$.

3. $-8(\mathbf{e} + \mathbf{f}) = (-8c, -8d) + (-8g, -8h)$

   Just remember to multiply each component by −8.

4. $-4(\mathbf{r} - \mathbf{s}) = (-24, -36) + (-32, 40)$

   Just remember to multiply each component by −4, paying careful attention to the signs.

5. $\mathbf{q} \cdot \mathbf{r} = 0$

   By formula: $\mathbf{q} \cdot \mathbf{r} = x_q x_r + y_q y_r$.

   Substitute: $\mathbf{q} \cdot \mathbf{r} = (6)(-5) + (5)(6)$

   Multiply: $\mathbf{q} \cdot \mathbf{r} = -30 + 30$

   And add: $\mathbf{q} \cdot \mathbf{r} = 0$.

6. $\mathbf{m} \cdot \mathbf{n} = 62$

    By formula: $\mathbf{m} \cdot \mathbf{n} = x_m x_n + y_m y_n$

    Substitute: $\mathbf{m} \cdot \mathbf{n} = (4)(5) + (-6)(-7)$

    Multiply: $\mathbf{m} \cdot \mathbf{n} = 20 + 42$

    And add: $\mathbf{m} \cdot \mathbf{n} = 62$

Chapter Problems and Solutions

Problems

Solve these problems for more practice applying the skills from this chapter. Worked out solutions follow problems.

1. Draw a vector **n** to represent a wind blowing north at 35 miles per hour for one hour and a vector **m** representing a motorist traveling west who has gone 35 miles.

2. Draw a vector representing a weight of 42 ounces.

3. Draw the resultant of vectors $\overrightarrow{AB}$ and $\overrightarrow{CD}$ as shown in the following figure.

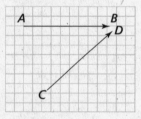

Vectors for problem 3.

4. Draw the resultant of vectors **m** and **n** as shown in the following figure.

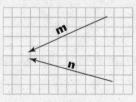

Vectors for problem 4.

5. Draw the resultant of vectors **v** and **u** as shown in the following figure.

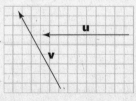

Vectors for problem 5.

6. Solve the parallelogram of forces in the following figure.

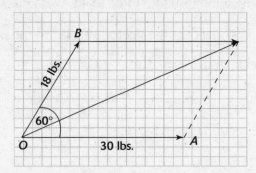

Diagram for problem 6.

7. A force of 24 kg and a force of 32 kg are acting on a body at an angle of 90°. What are the magnitude and angle of the resultant with respect to the 32 kg force?

8. An airplane is trying to fly due east at 400 mph. Because of a wind from the south, the plane's actual path is 25° north of east. Assuming the wind to be blowing at a steady pace, what is its magnitude, to the nearest whole mph?

9. A force of 10 lbs. and a force of 24 lbs. are acting on a body at an angle of 90°. What are the magnitude and angle of the resultant with respect to the 24-lb. force?

10. If the endpoints of vector $\overrightarrow{AB}$ are A (−4, −6) and B (4, 6), what are the coordinates of standard vector $\overrightarrow{OP}$ if $\overrightarrow{OP} = \overrightarrow{AB}$?

11. If the endpoints of vector $\overrightarrow{CD}$ are C (−5, −7) and D (4, 4), what are the coordinates of standard vector $\overrightarrow{OM}$ if $\overrightarrow{OM} = \overrightarrow{CD}$?

12. If the endpoints of vector $\overrightarrow{AB}$ are A (2, −5) and B (12, 7), what are the coordinates of standard vector $\overrightarrow{OP}$ if $\overrightarrow{OP} = \overrightarrow{AB}$?

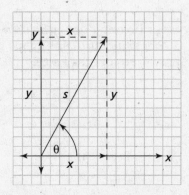

Model for problems 13 through 16.

Refer to the preceding figure as the model for problems 13–16.

13. Find the rectangular components of the vector 16∠30° to the nearest integer.

14. The rate and angle with the bank of a boat crossing a stream in miles per hour is 15∠70°. What are the cross-stream and current components of the boat's velocity?

15. The vertical component of a force is three times the magnitude of the horizontal component. Find the angle the resultant force makes with the smaller component.

16. Find the rectangular components of the vector 20∠36° to the nearest integer.

17. Find the magnitude of vector **h** = (−6, 10).

18. Vector **f** = (8, −3); vector **g** = (*l*, *m*). **f** = **g**. Find the value of *m*.

19. Find the magnitude of vector **m** = (−15, −36).

20. Multiply **v** (−5, 8) by 7.

21. If **r** = (*c*, *d*) and **s** = (*g*, *h*), find the product of 11 and **r** + **s**.

22. If **m** = (6, −9) and **n** = **(8, −7),** find the product of −8 and **m** − **n**.

23. Multiply **v** (−3, 6) by **u** (−5, 7).

24. If **n** = (*a*, *b*) and **o** = (*f*, *g*), find the product **n** · **o**.

25. If **m** = (5, −7) and **n** = (−6, 9), find the dot product of **m** and **n**.

Answers and Solutions

1. **Answer:**

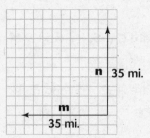

It is conventional to draw north as pointing upward and west to the left of north. It is also conventional to draw them as connected at their origins, although that was not necessary. It is, however, necessary to show them of equal size.

2. **Answer:**

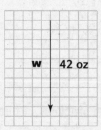

Weight always acts downward. Remember that. It's the pull of the force of gravity on a mass.

3. ***Answer:***

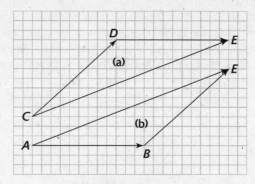

Two solutions to problem 3.

First, draw $\overrightarrow{CD}$. Then, at the tip of $\overrightarrow{CD}$ place $\overrightarrow{DE}$, the equivalent vector of $\overrightarrow{AB}$. Finally, connect the initial point of $\overrightarrow{CD}$ to the terminal point of $\overrightarrow{DE}$ to get resultant $\overrightarrow{CE}$. You could also have started with $\overrightarrow{AB}$ and then placed $\overrightarrow{BE}$, as in the figure. The resultant would have been $\overrightarrow{AE}$.

4. ***Answer:***

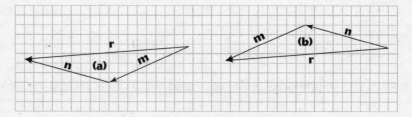

Two solutions to problem 4.

Bring the tip of vector **m** to the tail of **n** (a) or the other way around (b). In either case, connect the initial point of the first vector to the terminal point of the second to draw the resultant, **r.**

5. ***Answer:***

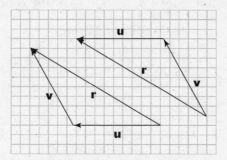

Two solutions to problem 5.

Bring the tip of vector **v** to the tail of **u** or the other way around. In either case, connect the initial point of the first vector to the terminal point of the second to draw the resultant, **r.**

6. ***Answer:*** 42 lbs. at $\approx 21.8°$.

Remember that the Law of Cosines can be used with any general triangle. The 60° angle between the vectors is not usable, since it is not contained wholly within either triangle. However, consecutive angles of a parallelogram are supplementary, so that means m∠A = 120°. 180° − 120° gives you a reference angle of 60°, and cosine is negative in quadrant II.

State the formula: $\cos A = \dfrac{b^2 + c^2 - a^2}{2bc}$

Substitute: $-\dfrac{1}{2} = \dfrac{18^2 + 30^2 - a^2}{18 \cdot 30 \cdot 2}$

Multiply: $-\dfrac{1}{2} = \dfrac{324 + 900 - a^2}{18 \cdot 60}$

Combine like terms: $-\dfrac{1}{2} = \dfrac{1224 - a^2}{1080}$

Cross multiply: $2(1224 - a^2) = -1 \cdot 1080$

Divde both sides by 2: $1224 - a^2 = -540$

Collect terms: $-a^2 = -540 - 1224$

Add; multiply by −1: $a^2 = 1764$

$\sqrt{\text{both sides:}}$ $a = 42$ lbs.

The resultant is 42 lbs. in magnitude. Next, you need the ∠COA. To find it, use the Law of Sines.

State the formula: $\dfrac{a}{\sin A} = \dfrac{\text{side opposite} \angle COA}{\sin \angle COA}$

Substitute: $\dfrac{42}{\sin 120°} = \dfrac{18}{\sin \angle COA}$

Cross multiply: $42 \sin \angle COA = 18 \sin 120°$

The quadrant II reference angle for 120° is 60°, and sine is positive in quadrant II, so:

$$42 \sin \angle COA = 18 \sin 60°$$

Substitute and divide both sides by 42: $\sin \angle COA \dfrac{18 \cdot 0.867}{42} = \dfrac{15.606}{42}$

Divide by 42: $\sin \angle COA = 0.372$

Find arcsin or $\sin^{-1} 0.372$: $\angle COA \approx 21.8°$

7.   **_Answer:_** 40 kg ≈ 36.9°

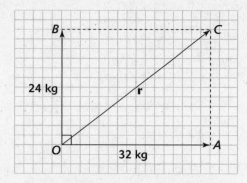

Parallelogram of forces for problem 7.

Always start by drawing a diagram as shown in the preceding figure. $\overrightarrow{OB}$ represents the 32 kg force, and $\overrightarrow{OA}$ represents the 24 kg force. O is the object upon which the forces are acting, and $\overrightarrow{OC}$ is the resultant force. Notice the corners of this parallelogram are right angles, so the parallelogram of forces is, in fact, a rectangle of forces. That means you can find $\overrightarrow{OC}$ by using the Pythagorean theorem.

State the equation: $c^2 = a^2 + b^2$

Substitute: $c^2 = 24^2 + 32^2$

Square the appropriate numbers: $c^2 = 576 + 1024$

Add: $c^2 = 1600$

Take the square root of both sides: $c = 40$

So the resultant is 40 kg in magnitude. To find its angle with respect to the 32 kg force, you must find $\angle COB$.

You know that the tangent of $<COB = \frac{24}{32} = \frac{3}{4} = 0.75$.

Look that up in the table or on your calculator and you'll find $\tan^{-1} 0.75 \approx 36.9°$.

8.   **_Answer:_** 186.4 mph

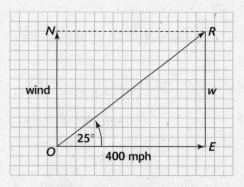

Plane is blown north of east.

Start with the diagram in which w is the equivalent vector of the wind, $\overrightarrow{ON}$. With respect to the angle at O, the relevant relationship is tangent.

State the equation: $\tan\theta = \dfrac{\text{opposite}}{\text{adjacent}}$

Substitute: $\tan 25° = \tan 25° = \dfrac{w}{400}$

Cross multiply: $w = 400\tan 25°$

Find $\tan 15°$: $w = 400(0.466).$

Multiply: $w = 186.4$ mphmph.

9.   **Answer:** 26 lbs., $\approx 23°$

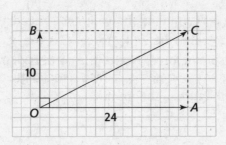

Diagram for solving problem 9.

Start by drawing a diagram as in the preceding figure. $\overrightarrow{OB}$ represents the 10 lbs. force, and $\overrightarrow{OA}$ represents the 24 lbs. force. O is the object upon which the forces are acting, and $\overrightarrow{OC}$ is the resultant force. Notice the corners of this parallelogram are right angles, so the parallelogram of forces is, in fact, a rectangle of forces. That means you can find OC through the magic of the Pythagorean theorem.

State the equation: $c^2 = a^2 + b^2$

Substitute: $c^2 = 10^2 + 24^2$

Square the appropriate numbers: $c^2 = 100 + 576$

Add: $c^2 = 676$

Take the square root of both sides: $c = 26$

The resultant is 26 lbs. in magnitude. To find its angle with respect to the 24 lb. force, you must find $\angle COB$.

You know that the tangent of $\angle COB = \dfrac{10}{24} = \dfrac{5}{12} = 0.417.$

Look that up in the table or on your calculator, and you'll find $\tan^{-1} 0.417 \approx 23°$.

10. **Answer:** (8, 12)   Remember, if the coordinates of point A are (x_a, y_a) and the coordinates of point B are (x_b, y_b), then the coordinates of point P are $(x_b - x_a, y_b - y_a)$.

    Substitute: $(4 - -4, 6 - -6)$

    Add: (8, 12)

    That, of course, is the terminal point of $\overrightarrow{OP}$, its initial point being at the origin.

11. **Answer:** (9, 11)   Since the coordinates of point A are (x_a, y_a) and the coordinates of point B are (x_b, y_b), then the coordinates of point M are $(x_b - x_a, y_b - y_a)$.

    Substitute: $(4 - -5, 4 - -7)$

    Add: (9, 11)

12. **Answer:** (10, 12)   Remember, if the coordinates of point A are (x_a, y_a) and the coordinates of point B are (x_b, y_b), then the coordinates of point P are $(x_b - x_a, y_b - y_a)$.

    Substitute: $(12 - 2, 7 - -5)$

    Add: (10, 12)

13. **Answer:** $x = 14$, $y = 8$

    Start with sine: $\sin 30° = \dfrac{\text{opposite}}{\text{hypotenuse}}$

    Substitute: $\dfrac{1}{2} = \dfrac{y}{16}$

    Cross multiply: $2y = 16$

    Divide both sides by 2: $y = 8$

    Next, use either tangent or cosine: $\tan 30° = \dfrac{\text{opposite}}{\text{adjacent}}$

    Substitute remembering that $\tan 30° = \dfrac{1}{\sqrt{3}}$: $\dfrac{1}{\sqrt{3}} = \dfrac{8}{x}$

    Cross multiply: $x = 8\sqrt{3}$

    Leave in radical form or multiply: $x = 8\sqrt{3}$ or 14

14. **Answer:** Current is about 5 mph, cross-stream rate is about 14 mph. Since the angle with the bank is 70°, you can safely conclude from reference to the figure on page 189 that the x component is the downstream one, that is, the current, and that the cross stream rate is the y component.

    Start with sine: $\sin 70° = \dfrac{\text{opposite}}{\text{hypotenuse}}$

    Substitute: $\dfrac{0.94}{1} = \dfrac{y}{15}$

    Cross multiply: $y = 15 \cdot 0.94$

Multiply and round: $\qquad\qquad\qquad\qquad\qquad\qquad$ $y = 15 \cdot 0.94 \approx 14$

Next, use cosine, since the y-value is not exact: $\quad \cos 70° = \dfrac{\text{adjacent}}{\text{hypotenuse}}$

Substitute: $\qquad\qquad\qquad\qquad\qquad\qquad$ $\dfrac{0.342}{1} = \dfrac{x}{15}$

Cross multiply: $\qquad\qquad\qquad\qquad\qquad\qquad$ $x = 15 \cdot 0.342$

Multiply and round: $\qquad\qquad\qquad\qquad\qquad\quad$ $x = 5.13 \approx 5$

15. **Answer:** $\approx 72°$ Let the angle be ϕ, the horizontal component be h and the vertical component be $3h$.

Use the tangent function, since: $\qquad\qquad$ $\tan\phi = \dfrac{\text{opposite}}{\text{adjacent}}$

Substitute: $\qquad\qquad\qquad\qquad\qquad\qquad$ $\tan\phi = \dfrac{3h}{h}$

But the hs cancel each other, so: $\qquad\quad$ $\tan\phi = \dfrac{3\cancel{h}}{1\cancel{h}} = 3$

Now find the angle whose tangent is 3: $\quad \tan\phi = 71.57 \approx 72°$

16. **Answer:** $x \approx 16,\ y \approx 12$

Start with sine: $\qquad\qquad\qquad\qquad\qquad$ $\sin 36° = \dfrac{\text{opposite}}{\text{hypotenuse}}$

Substitute: $\qquad\qquad\qquad\qquad\qquad\qquad$ $\dfrac{0.588}{1} = \dfrac{y}{20}$

Cross multiply: $\qquad\qquad\qquad\qquad\qquad$ $y = 20(0.588) = 11.76 \approx 12$

Next, use cosine, since y is inexact: $\quad \cos 36° = \dfrac{\text{adjacent}}{\text{hypotenuse}}$

Substitute: $\qquad\qquad\qquad\qquad\qquad\qquad$ $\dfrac{0.809}{1} = \dfrac{x}{20}$

Cross multiply: $\qquad\qquad\qquad\qquad\qquad$ $x = 20(0.809)$

Multiply: $\qquad\qquad\qquad\qquad\qquad\qquad$ $x = 16.18 \approx 16$

17. **Answer:** $2\sqrt{34}$ Remember that a vector has a horizontal and a vertical component, which are −6 and 10, respectively. The resultant may be represented as the hypotenuse of a right triangle.

You can use the Pythagorean theorem or the distance formula to solve.

Write the formula: $\qquad\qquad\qquad\qquad$ $|h| = \sqrt{a^2 + b^2}$

Substitute: $\qquad\qquad\qquad\qquad\qquad\qquad$ $|h| = \sqrt{(-6)^2 + 10^2}$

Square: $\qquad\qquad\qquad\qquad\qquad\qquad\quad$ $|h| = \sqrt{36 + 100}$

Combine terms: $\qquad\qquad\qquad\qquad\quad$ $|h| = \sqrt{136}$

Simplify the radical by dividing by 4: $\quad |h| = 2\sqrt{34}$

18. ***Answer:*** −3   And the value of l is 8. The statement **f = g** means they have the same (or identical) components. They are equal vectors. That means they have the same horizontal and vertical components.

19. ***Answer:*** 39   Remember that a vector has both a horizontal and a vertical component, which are −15 and −36, respectively. The resultant may be represented as the hypotenuse of a right triangle. You can use the Pythagorean theorem or the distance formula to solve.

    Write the formula: $|m| = \sqrt{a^2 + b^2}$

    Substitute: $|m| = \sqrt{(-15)^2 + (-36)^2}$

    Square: $|m| = \sqrt{225 + 1296}$

    Combine terms: $|m| = \sqrt{1521}$

    Solve the radical: $|m| = 39$

20. ***Answer:*** 7**v** = (−35, 56)   Simply multiply both components by the scalar, 7.

21. ***Answer:*** 11(**r** + **s**) = (11c, 11d) + (11g, 11h)   Just remember to multiply each component by 11.

22. ***Answer:*** −8(**m** − **n**) = (−48, 72) − (−64, 56)   Just remember to multiply each component by −7, paying careful attention to the signs.

23. ***Answer:*** **v** · **u** = 57

    By formula: $\mathbf{v} \cdot \mathbf{u} = x_u x_v + y_u y_v.$

    Substitute: $\mathbf{v} \cdot \mathbf{u} = -3 \cdot -5 + 6 \cdot 7$

    Multiply: $\mathbf{v} \cdot \mathbf{u} = 15 + 42$

    And add: $\mathbf{v} \cdot \mathbf{u} = 57$

24. ***Answer:*** **n** · **o** = $a\hat{f} + bg$

    By formula: $\mathbf{n} \cdot \mathbf{o} = x_n x_o + y_n y_o$

    Substitute: $\mathbf{n} \cdot \mathbf{o} = af + bg$

    And that's as far as you can go.

25. ***Answer:*** **m** · **n** = −93

    By formula: $\mathbf{m} \cdot \mathbf{n} = x_m x_n + y_m y_n$

    Substitute: $\mathbf{m} \cdot \mathbf{n} = (5)(-6) + (-7)(9)$

    Multiply: $\mathbf{m} \cdot \mathbf{n} = -30 + -63$

    And add: $\mathbf{m} \cdot \mathbf{n} = -93$

Supplemental Chapter Problems

Solve these problems for even more practice applying the skills from this chapter. The Answer section will direct you to where you need to review.

Problems

1. Draw a vector **s** to represent a wind blowing south at 100 miles per hour for one hour and a vector **p** representing a plane traveling east that has gone 300 miles.

2. Draw a vector representing a weight of 230 pounds.

3. Draw the resultant of vectors $\vec{AB}$ and $\vec{CD}$ as shown in the following figure.

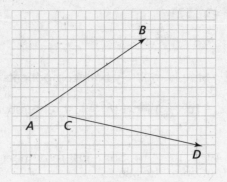

Vectors for problem 3.

4. Draw the resultant of vectors **v** and **u** as shown in the following figure.

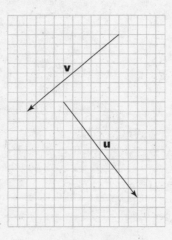

Vectors for problem 4.

5. Draw the resultant of vectors **u** and **v** as shown in the following figure.

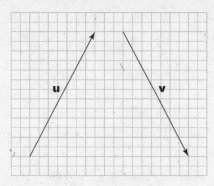

Vectors for problem 5.

6. Solve the parallelogram of forces in the following figure.

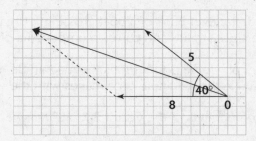

Diagram for problem 6.

7. A force of 14 lbs. and a force of 12 lbs. are acting on a body at an angle of 90°. What are the magnitude and angle of the resultant with respect to the 14 lb. force?

8. An airplane is trying to fly due west at 600 mph. Because of a wind from the north, the plane's actual path is 10° south of west. Assuming the wind to be blowing at a steady pace, what is its magnitude to the nearest whole mph?

9. What is the magnitude of the speed of the plane in Problem 8?

10. If the endpoints of vector $\overrightarrow{AB}$ are A (−3, −7) and B (2, 5), what are the coordinates of standard vector $\overrightarrow{OP}$ if $\overrightarrow{OP} = \overrightarrow{AB}$?

11. If the endpoints of vector $\overrightarrow{CD}$ are C (−6, −9) and D (2, 4), what are the coordinates of standard vector $\overrightarrow{OM}$ if $\overrightarrow{OM} = \overrightarrow{CD}$?

12. If the endpoints of vector $\overrightarrow{AB}$ are A (3, 6) and B (13, 14), what are the coordinates of standard vector $\overrightarrow{OP}$ if $\overrightarrow{OP} = \overrightarrow{AB}$?

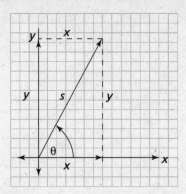

Model for problems 13 through 16.

Refer to the preceding figure as the model for Problems 13–16.

13. Find the rectangular components of the vector 30∠30° to the nearest integer.

14. The rate and angle with the river bank of a boat crossing a stream in miles per hour is 20∠70°. What are the cross stream and current components of the boat's velocity?

15. The vertical component of a force is four times the magnitude of the horizontal component. Find the angle the resultant force makes with the smaller component.

16. Find the rectangular components of the vector 30∠45° (answer may be left in radical form).

17. Find the magnitude of vector **h** = (−5, 9)

18. Vector **b** = (6, −9); vector **c** = (p, q); **b** = **c.** Find the value of p.

19. Find the magnitude of vector **m** = (−12, −16)

20. Multiply **v** (−8, 9) by 10.

21. If **r** = (m, n) and **s** = (p, q), find the product of 9 and **r** + **s.**

22. If **m** = (7, −8) and **n** = (6, −9), find the product of −7 and **m** − **n.**

23. Multiply **v** (5, −4) by **u** (−8, 7).

24. If **r** = (c, d) and **s** = (a, b), find the product **r** · **s.**

25. If **p** = (5, −8) and **q** = (−7, 9), find the dot product of **p** and **q.**

Answers

1.

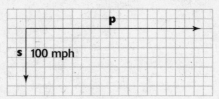

(Vectors versus Scalars, p. 167)

2.

(Vectors versus Scalars, p. 167)

3.

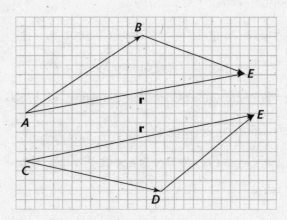

(Vector Addition Triangle/The Tip-Tail Rule, p. 169)

4.

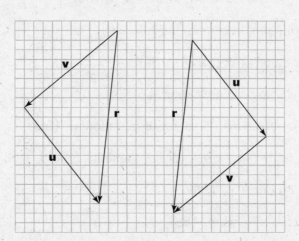

(Vector Addition Triangle/The Tip-Tail Rule, p. 169)

5.

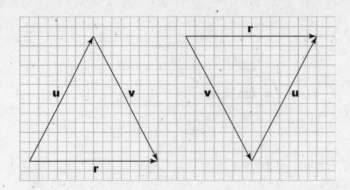

(Vector Addition Triangle/The Tip-Tail Rule, p. 169)

6. 12.3 lbs. at ≈ 15.14°   (Parallelogram of Forces, p. 172)

7. 18.44 lbs. ≈ 40.6°   (Parallelogram of Forces, p. 172)

8. 105.6 mph   (Parallelogram of Forces, p. 172)

9. 609 mph   (Parallelogram of Forces, p. 172)

10. (5, 12)   (Vectors in the Rectangular Coordinate System, p. 178)

11. (8, 13)   (Vectors in the Rectangular Coordinate System, p. 178)

12. (10, 8)   (Vectors in the Rectangular Coordinate System, p. 178)

13. $x ≈ 26, y = 15$ (Resolution of Vectors, p. 180)

14. Current is about 7 mph, cross-stream rate is about 19 mph   (Resolution of Vectors, p. 180)

15. ≈ 76°   (Resolution of Vectors, p. 180)

16. $\left(15\sqrt{2}, 15\sqrt{2}\right)$ or (21.213, 21.213)   (Resolution of Vectors, p. 180)

17. $\sqrt{106}$ or 10.3   (Algebraic Addition of Vectors, p. 183)

18. 6   (Algebraic Addition of Vectors, p. 183)

19. 20   (Algebraic Addition of Vectors, p. 183)

20. $10\mathbf{v} = (-80, 90)$ (Scalar Multiplication, p. 185)

21. $9(\mathbf{r} + \mathbf{s}) = (9m, 9n) + (9p, 9q)$ (Scalar Multiplication, p. 185)

22. $-7(\mathbf{m} - \mathbf{n}) = (-49, 56) - (-42, 63)$ (Scalar Multiplication, p. 185)

23. $\mathbf{v} \cdot \mathbf{u} = -68$ (Dot Products, p. 185)

24. $\mathbf{r} \cdot \mathbf{s} = ac + bd$ (Dot Products, p. 185)

25. $\mathbf{p} \cdot \mathbf{q} = -107$ (Dot Products, p. 185)

Chapter 6
Polar Coordinates and Complex Numbers

Many different forms of measure are common today, as are many different coordinate systems. When working on a flat surface, it is common to use rectangular (a.k.a., Cartesian) coordinates. When measuring a spherical surface like that of the earth, it is common to use degrees, minutes, and seconds of latitude and longitude. Under certain conditions, especially those involving periodic functions, polar coordinates work best. The polar coordinate system is second only to Cartesian coordinates in the universality of its usage.

Polar Coordinates

The **polar coordinate system** consists of a fixed point, usually designated O and known as the **pole,** or **origin.** Extending to the right of the pole and horizontal is a ray known as the **polar axis.** We can locate any point, P, by specifying an angle, θ, with respect to the polar axis and a distance from the origin, r, as in the following figure. The ordered pair, (r, θ), names the **polar coordinates** of point P.

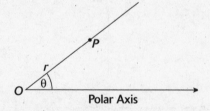

Polar axis, pole, angleθ, and point P.

If $\angle\theta$ is measured counterclockwise from the polar axis, then it is a positive angle. If it were measured clockwise, it would be a negative angle. The angle in the preceding figure would normally be interpreted to be positive angle θ, but it could also be expressed as $-(360° - \theta)$, not that it would make a a lot of sense to do so. For r to be positive, P must be on the north side of the polar axis. On the south side of the pole, the value of r is negative (see the following figure).

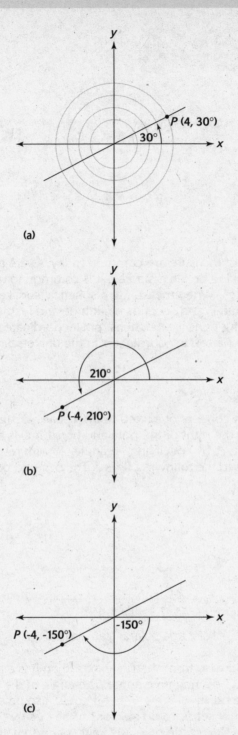

Polar forms of three coterminal angles.

Note that (a) has a positive r and a positive θ. (b) has a negative r and a positive θ. In (c), both r and θ are negative.

Example Problems

These problems show the answers and solutions.

Write the polar coordinates of each of the following points, P. Express each as an ordered pair.

1.

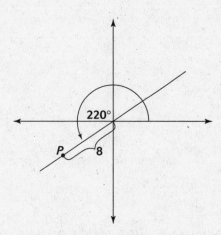

Diagram for Problem 1.

answer: (–8, 220°)

P is below the axis, so r is negative. θ is rotated counterclockwise, and so it is positive.

2.

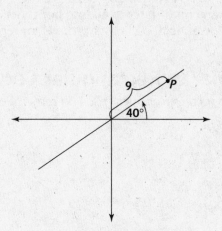

Diagram for Problem 2.

answer: (9, 40°)

P is above the axis, so r is positive. θ is rotated counterclockwise, and so it is positive.

3.

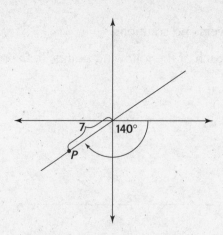

Diagram for Problem 3.

answer: **(–7, –140°)**

P is below the axis, so r is negative. θ is rotated clockwise, and so it is negative.

Converting between Polar and Rectangular Coordinates

It will help to convert from polar to rectangular coordinates if you remember that the distance from the pole of a point on a polar graph may be resolved into horizontal and vertical coordinates, just as you did with vectors. In fact, polar notation is useful for vector representation as well.

Converting from Polar to Rectangular Coordinates

Converting polar coordinates to rectangular ones relies on remembering ratios from the unit circle.

To get x, use cosine: $\frac{x}{r} = \cos\theta$

Cross multiplying gets: $x = r\cos\theta$

Similarly for y: $\frac{y}{r} = \sin\theta$

Cross multiply to get: $y = r\cos\theta$

Converting from Rectangular to Polar Coordinates

To convert from rectangualr to polar coordinates, the procedure is similar. Starting with the Pythagorean theorem, you get the Distance Formula: $r = \sqrt{x^2 + y^2}$.

Then use cosine θ or sine θ or tangent θ to find the angle. Having done that, you'll have the polar coordinates, (r, θ).

Example Problems

1. Convert $P(6, 8)$ to polar coordinates.

   answer: **(10, 53°)**

   First, use the formula: $r^2 = x^2 + y^2$

   Solve that for r: $r = \sqrt{x^2 + y^2}$

   Substitute: $r = \sqrt{6^2 + 8^2}$

   Square terms indicated: $r = \sqrt{36 + 64}$

   Add and take the square root: $r = \sqrt{100} = 10$

   Next, pick a trigonometric ratio: $\sin\theta = \dfrac{y}{r}$

   Substitute: $\sin\theta = \dfrac{8}{10} = 0.8$

   Use $\sin^{-1}\theta$ to find the angle: $\sin^{-1}\theta = 0.8; \ \theta \approx 53°$

   The polar coordinates for $P(6,8)$: $(r, \theta) = (10, 53°)$

2. Convert $P(12, 30°)$ to rectangular coordinates:

   answer: **$P(10.4, 6)$**

   First, find x using cosine: $\dfrac{x}{r} = \cos\theta$

   Cross multiplying gets: $x = r\cos\theta$

   Substitute: $x = 12\cos30°$

   Solve: $x = 12(0.866) \approx 10.4$

   Similarly for y: $\dfrac{y}{r} = \sin\theta$

   Cross multiply to get: $y = r\sin\theta$

   Substitute: $y = 12\sin30°$

   Solve: $y = 12(0.5) = 6$

   Therefore, $P(12, 30°)$ has rectangular coordinates $P(10.4, 6)$.

3. Convert $P(5, 9)$ to polar coordinates.

    answer: $(\sqrt{106}, 61°)$

    First, use the formula: $\qquad\qquad$ $r^2 = x^2 + y^2$

    Solve that for r: $\qquad\qquad\qquad$ $r = \sqrt{x^2 + y^2}$

    Substitute: $\qquad\qquad\qquad\quad$ $r = \sqrt{5^2 + 9^2}$

    Square terms indicated: $\qquad\quad$ $r = \sqrt{25 + 81}$

    Add and take the square root: $\quad$ $r = \sqrt{106}$

    Next, pick a trigonometric ratio: $\quad$ $\tan\theta = \dfrac{y}{x}$

    Substitute: $\qquad\qquad\qquad\qquad$ $\tan\theta = \dfrac{9}{5} = 1.8$

    Use $\tan^{-1}\theta$ to find the angle: $\quad$ $\tan^{-1}\theta = 1.8;\ \theta \approx 61°$

    The polar coordinates for $P(5, 9)$: $\quad (r, \theta) = (\sqrt{106}, 61°)$

4. Convert $P(8, 60°)$ to rectangular coordinates.

    answer: $P(4, 7)$

    First, find x using cosine: $\quad \dfrac{x}{r} = \cos\theta$

    Cross multiplying gets: $\qquad x = r\cos\theta$

    Substitute: $\qquad\qquad\qquad x = 8\cos 60°$

    Solve: $\qquad\qquad\qquad\qquad x = 8(0.5) = 4$

    Similarly for y: $\qquad\qquad \dfrac{y}{r} = \sin\theta$

    Cross multiply to get: $\qquad y = r\sin\theta$

    Substitute: $\qquad\qquad\qquad y = 8\sin 60°$

    Solve: $\qquad\qquad\qquad\qquad y = 8(0.866) = 6.928 \approx 7$

    Therefore, $P(8, 60°)$ has rectangular coordinates $P(4, 7)$.

Work Problems

Use these problems to give yourself additional practice.

1. Write the polar coordinates of P.

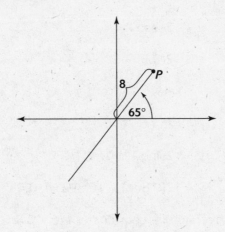

Diagram for Problem 1.

2. Convert $P(10, 40°)$ to rectangular coordinates.

3. Convert $P(12, 60°)$ to rectangular coordinates.

4. Convert $P(9, 12)$ to polar coordinates.

5. Convert $P(6, 10)$ to polar coordinates.

Worked Solutions

1. **(8, 65°)** P is above the axis, so r is positive. θ is rotated counterclockwise, and so is positive.

2. **P(7.7, 6.4)**

    First, find x using cosine: $\frac{x}{r} = \cos\theta$

    Cross multiplying gets: $x = r\cos\theta$

    Substitute: $x = 10\cos 40°$

    Solve: $x = 10(0.766) \approx 7.7$

    Similarly for y: $\frac{y}{r} = \sin\theta$

    Cross multiply to get: $y = r\sin\theta$

    Substitute: $y = 10\sin 40°$

    Solve: $y = 10(0.643) \approx 6.4$

    Therefore, $P(10, 40°)$ has rectangular coordinates $P(7.7, 6.4)$.

3. **P(6, 10.4)**

   First, find x using cosine: $\quad \frac{x}{r} = \cos\theta$

   Cross multiplying gets: $\quad x = r\cos\theta$

   Substitute: $\quad x = 12\cos60°$

   Solve: $\quad x = 12(0.5) = 6$

   Similarly for y: $\quad \frac{y}{r} = \sin\theta$

   Cross multiply to get: $\quad y = r\sin\theta$

   Substitute: $\quad y = 12\sin60°$

   Solve: $\quad y = 12(0.866) = 10.392 \approx 10.4$

   Therefore, $P(12, 60°)$ has rectangular coordinates $P(6, 10.4)$

4. **(15, 53°)**

   First, use the formula: $\quad r^2 = x^2 + y^2$

   Solve that for r: $\quad r = \sqrt{x^2 + y^2}$

   Substitute: $\quad r = \sqrt{9^2 + 12^2}$

   Square terms indicated: $\quad r = \sqrt{81 + 144}$

   Add and take the square root: $\quad r = \sqrt{225} = 15$

   Next, pick a trigonometric ratio: $\quad \sin\theta = \frac{y}{r}$

   Substitute: $\quad \sin\theta = \frac{12}{15} = 0.8$

   Use $\sin^{-1}\theta$ to find the angle: $\quad \sin^{-1}\theta = 0.8;\ \theta \approx 53°$

   The polar coordinates for $P(9,12)$: $\quad (r,\ \theta) = (15, 53°)$

5. **$(2\sqrt{34}, 59°)$**

   First, use the formula: $\quad r^2 = x^2 + y^2$

   Solve that for r: $\quad r = \sqrt{x^2 + y^2}$

   Substitute: $\quad r = \sqrt{6^2 + 10^2}$

   Square terms indicated: $\quad r = \sqrt{36 + 100}$

   Add and take the square root: $\quad r = \sqrt{136} = 2\sqrt{34}$

   Next, pick a trigonometric ratio: $\quad \tan\theta = \frac{y}{x}$

Substitute: $\tan\theta = \dfrac{10}{6} = 1.667$

Use $\tan^{-1}\theta$ to find the angle:　$\tan^{-1}\theta = 1.667$; $\theta \approx 59°$

The polar coordinates for $P(6, 10)$:　$(r, \theta) = (2\sqrt{34}, 59°)$

Some Showy Polar Graphs

Some polar graphs are very distinctive and are presented here for your attention and enjoyment. Don't worry; you won't be expected to memorize (most of) them or draw them, unless you go on to an engineering course of study in college. In all the equations, *a* is a constant that determines the scale of the curve, that is, its size.

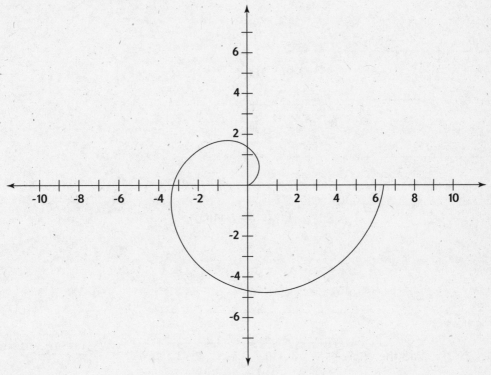

Archimedes Spiral $r \approx a\theta$.

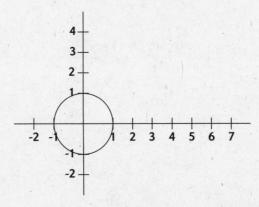

Circle: $r = a$.

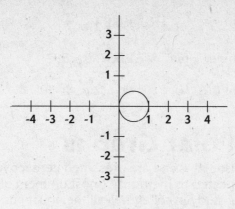

Circle: $r = a\cos\theta$.

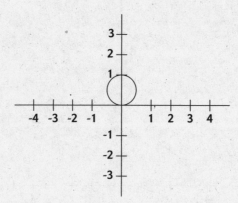

Circle: $r = a\sin\theta$.

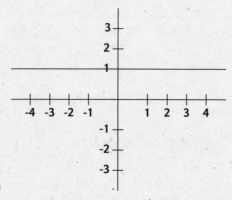

Horizontal line: $r = \dfrac{a}{\sin\theta}$.

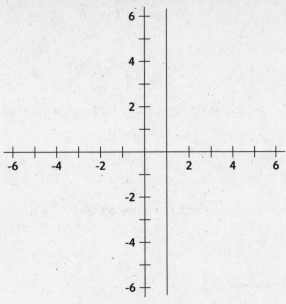

Vertical line: $r = \dfrac{a}{\cos\theta}$.

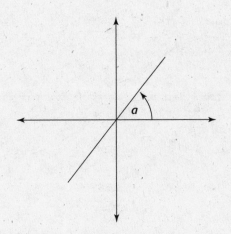

Line: $\theta = a$.

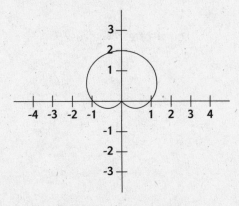

Cardoid: $r = a + a\sin\theta$.

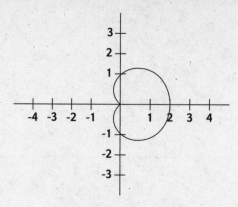

Cardoid: r = a + acosθ.

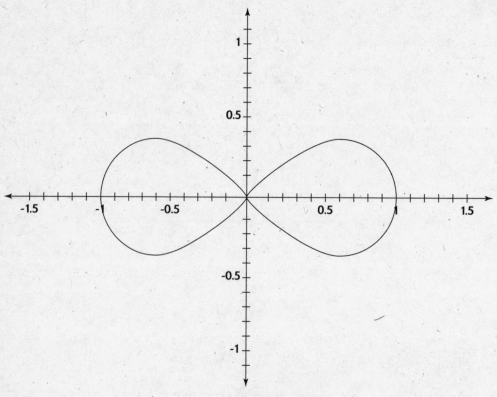

Lemniscate: r² = a²cos2θ.

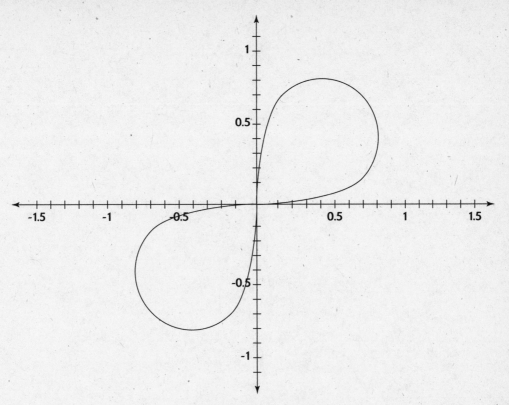

Lemniscate: $r^2 = a^2 \sin 2\theta$.

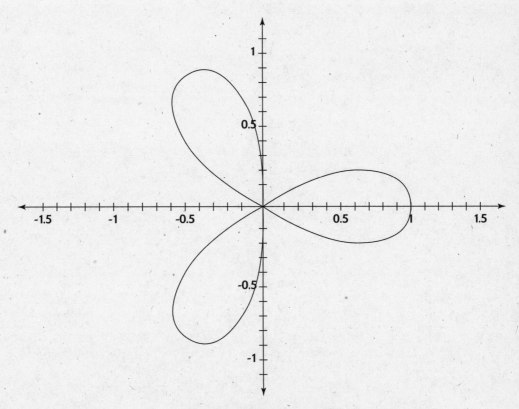

Three-leaved rose: $r = a\cos 3\theta$.

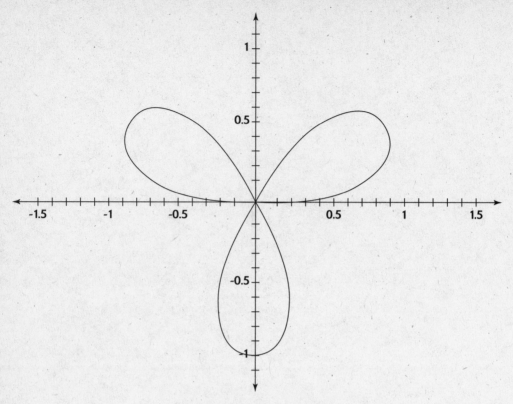

Three-leaved rose: r = asin3θ.

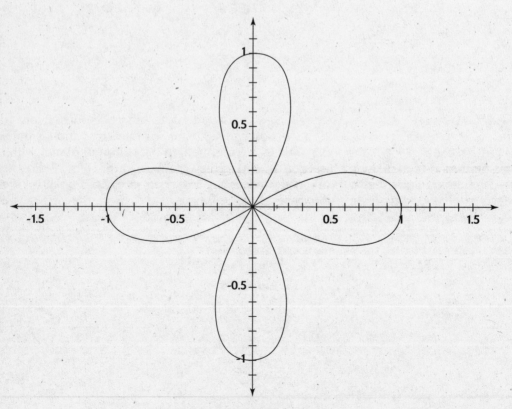

Four-leaved rose: r = acos2θ.

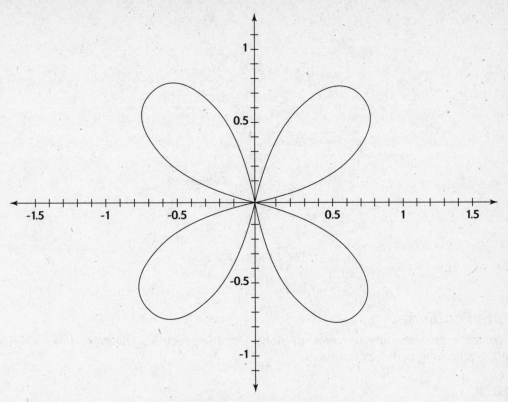

Four-leaved rose: r = asin2θ.

Now is that cool or what?!! By the way, *cardoid* means heartlike.

Plotting Complex Numbers on Rectangular Axes

Complex numbers are plottable on both rectangular coordinates and polar ones. Every complex number can be written in the form $a + bi$ in which a and b are real numbers, and i is the square root of negative 1 ($i^2 = -1$). Complex numbers are graphed on the **complex plane,** with real numbers figured in the horizontal direction and imaginary numbers in the vertical direction; that is to say, the x-axis is the real axis, and the y-axis is the imaginary one. Each point on the complex plane may be designated by an ordered pair, designated (a, b). Since the vertical axis is the imaginary one, the i in $a + bi$ need not be written. It is understood. It is customary in some circles to designate the axes in the complex plane by uppercase X and Y to distinguish them from the Cartesian plane axes, x and y.

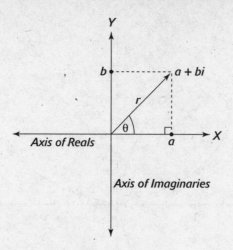

The complex plane.

Example Problems

These problems show the answers and solutions. Use the preceding diagram as a model for plotting the following complex numbers:

1. $4 + 3i$

   answer: **See the following figure**

   Both signs are positive, so it goes in quadrant I.

2. $-3 + 5i$

   answer: **See the following figure**

   A negative a coordinate and a positive b coordinate put the number in the second quadrant.

3. $-4 - 3i$

   answer: **See the following figure**

   $-4 - 3i$ could have been written as $-4 + -3i$. In either case there are two negative coordinates, so it falls in the third quadrant.

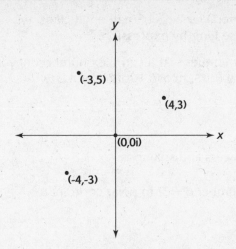

Answers to example problems.

Notice the coordinates of the origin are (0, 0i).

Plotting Complex Numbers on the Polar Axis

A glance back at the figure used as a model for the preceding set of example problems will show you that the polar representation of complex numbers was anticipated in that diagram, since rectangular axes, the angle theta, and the vector r were included. If you remember the trigonometric ratios, then you should recall that given a point, P, with rectangular coordinates (x, y), x could be represented as $r\cos\theta$, and y could be represented as $r\sin\theta$ (see the following figure).

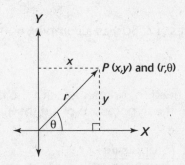

Point P and pertinent rectangular and polar information.

If the complex number is $x + yi$, then, based upon the representations above, $x + yi$ in polar form could be shown by the following equations:

$$x + yi = r\cos\theta + ir\sin\theta$$

Factor out an r and get : $$x + yi = r(\cos\theta + i\sin\theta)$$

The expression $r(\cos\theta + i\sin\theta)$ is called the **polar form** or **trigonometric form** of the complex number, $x + yi$. The number r is known as the **modulus** or **absolute** value of the complex number, and the angle θ (between the polar axis and the line drawn from the pole) is called the **amplitude** or **argument** of the complex number. By using c to stand for cosine and s to stand for sine, the expression $\cos\theta + i\sin\theta$ can be abbreviated $cis\theta$. You can then write, $x + yi = rcis\theta$.

This is a testimonial to the general laziness of mathematicians, who are always looking for a short way to write an otherwise lengthy expression.

To summarize this section, remember that a complex number may be represented by a point on the polar plane represented by (r, θ), $r(\cos\theta + i\sin\theta)$, or $r cis \theta$.

Example Problems

These problems show the answers and solutions.

1. Convert the complex number $6 + 3i$ to polar coordinates.

    answer: $(3\sqrt{5}, \approx 27°)$

    The 6 is the horizontal coordinate, and $3i$ is the vertical coordinate. To find r, use the Distance Formula version of the Pythagorean theorem.

    Formula: $r = \sqrt{x^2 + y^2}$

    Substitute: $r = \sqrt{6^2 + 3^2}$

    Square: $r = \sqrt{36 + 9}$

    Simplify: $r = \sqrt{45} = \sqrt{9 \cdot 5} = \sqrt{9} \cdot \sqrt{5} = 3\sqrt{5}$

    Use tangent or cosine to find the angle: $\tan\theta = \dfrac{3}{6} = 0.5$

    Find the angle with table or calculator: $\tan^{-1}\theta = 0.5$; $\theta = 26.57°$

    The angle is about 27°.

2. Express the polar coordinates $(12, 30°)$ as a complex number in the form $a + bi$.

    answer: $10.4 + 6i$

    Draw the picture first if you need to, or refer back to the last figure. Since the first of the ordered pair is equivalent to the hypotenuse of a right triangle, you can find the imaginary coordinate from the sine ratio:

    $$\sin\theta = \frac{\text{opposite}}{\text{hypotenuse}}$$

    Substitute: $\sin 30° = \dfrac{y}{12}$

    Cross multiply: $y = 12\sin 30°$

    Substitute and multiply: $y = 12 \cdot 0.5 = 6$

    So you can conclude the imaginary part of the solution is $6i$.

    Now for the real part, use cos or tan: $\cos\theta = \dfrac{\text{adjacent}}{\text{hypotenuse}}$

    Substitute: $\cos\theta = \dfrac{x}{12}$

    Cross multiply: $x = 12\cos 30°$

Substitute and multiply: $x = 12 \cdot 0.866 = 10.392 \approx 10.4$

So the complex number is $10.4 + 6i$.

3. Express the complex number $4 + 4i$ as expressions in the forms $r(\cos\theta + i\sin\theta)$ and $r\text{cis}\theta$.

answer: $4\sqrt{2}\ (\cos 45° + i\sin 45°),\ 4\sqrt{2}(cis45°)$

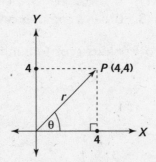

Model of Problem 3.

It should be obvious from the complex number with both rectangular coordinates of magnitude 4 that you are dealing with a parallelogram that is, in fact, a square. You may recall that in any square (or rhombus) the diagonal bisects the vertex angles. That makes $\theta = 45°$ (half of a right angle). Next, you need to find r, but you know $r = \sqrt{x^2 + y^2}$.

Substitute: $r = \sqrt{4^2 + 4^2}$

Square and combine: $r = \sqrt{16 + 16} = \sqrt{32}$

Simplify: $r = \sqrt{32} = \sqrt{16} \cdot \sqrt{2} = 4\sqrt{2}$

Therefore: $4 + 4i = 4\sqrt{2}(\cos 45° + i\sin 45°)$, or $4\sqrt{2}(cis45°)$

Work Problems

Use these problems to give yourself additional practice.

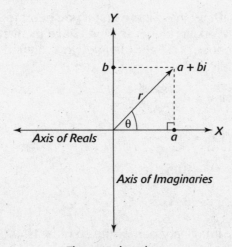

The complex plane.

Use the diagram as a model for plotting the following complex numbers:

1. $7 + 5i$

2. $-6 + 3i$

3. $-7 - 5i$

4. Express the polar coordinates $(8, 30°)$ as a complex number in the form $a + bi$.

5. Express the complex number $6 + 6i$ as expressions in the forms $r(\cos\theta + i\sin\theta)$ and $cis\theta$.

Worked Solutions

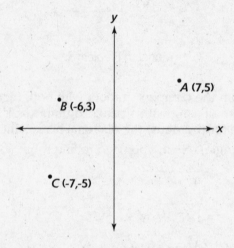

1. **See A on the preceding figure**   Both signs are positive, so it goes in quadrant I.

2. **See B on the preceding figure**   A negative a coordinate and a positive b coordinate put the number in the second quadrant.

3. **See C on the preceding figure**   A negative a coordinate and a negative b coordinate put the number in the third quadrant.

4. **$6.9 + 4i$, or $4\sqrt{3} + 4i$**   Draw the picture first if you need to, or refer back to the figure at the beginning of this Work Problems section. Since the first of the ordered pair is equivalent to the hypotenuse of a right triangle, you can find the imaginary coordinate from the sine ratio:

$$\sin\theta = \frac{\text{opposite}}{\text{hypotenuse}}$$

Substitute: $\sin 30° = \dfrac{y}{8}$

Cross multiply: $y = 8\sin 30°$

Substitute and multiply: $y = 8 \cdot 0.5 = 4$

So you can conclude the imaginary part of the solution is $4i$.

Now for the real part, use cos or tan: $\cos\theta = \dfrac{\text{adjacent}}{\text{hypotenuse}}$

Substitute: $\cos\theta = \dfrac{x}{8}$

Cross multiply: $x = 8\cos 30°$

Substitute and multiply: $x = 8 \cdot 0.866 = 6.928 \approx 6.9$

Or use the exact cos 30°: $x = 8 \cdot \dfrac{\sqrt{3}}{2} = 4\sqrt{3}$

So the complex number is $6.9 + 4i$, or $4\sqrt{3} + 4i$

5. **$6\sqrt{2}, 6\sqrt{2}$**   You should recognize from the complex number with both rectangular coordinates of magnitude 6 that you are dealing with a square. You may recall that in any square (or rhombus) the diagonal bisects the vertex angles. That makes $\theta = 45°$ (half of a right angle. Next, you need to find r, but you know:

$$r = \sqrt{x^2 + y^2}$$

Substitute: $r = \sqrt{6^2 + 6^2}$

Square and combine: $r = \sqrt{36 + 36} = \sqrt{72}$

Simplify: $r = \sqrt{72} = \sqrt{36} \cdot \sqrt{2} = 6\sqrt{2}$

Therefore: $6 + 6i = 6\sqrt{2}\,(\cos 45° + i\sin 45°)$, or $6\sqrt{2}\,(\text{cis}\,45°)$.

Conjugates of Complex Numbers

Two complex numbers are equal if and only if their real parts are equal and their imaginary parts are equal. That means that if $a + bi = x + yi$ then a must equal x and b must equal y. To further illustrate the principle, $3 + bi = x + 8i$ *if and only if* $x = 3$ and $b = 8$.

Every complex number, $a + bi$, has a conjugate, $a - bi$. In fact, $a + bi$ and $a - bi$ are conjugates of each other. The conjugate of $3 + 4i$ is $3 - 4i$; the conjugate of $7 - 6i$ is $7 + 6i$. You may assume that complex numbers follow the same rules for arithmetic properties as real numbers do. Using the conjugates $a + bi$ and $a - bi$ you get the following general rules for addition, subtraction, and multiplication of conjugates:

Sum: $(a + bi) + (a - bi) = (a + a) + (bi - bi) = 2a + 0 = 2a$

Difference: $(a + bi) - (a - bi) = (a - a) + (bi + bi) = 0a + 2bi = 2bi$

Product: $(a + bi)(a - bi) = a^2 + abi - abi - (bi)^2$

The abi terms add to 0, and $i^2 = -1$, so the above becomes $a^2 + -b^2(-1) = a^2 + b^2$.

Notice that the sum of two complex conjugates is a real number, but the difference is an imaginary one. The product will always be a positive real number, unless both a and b are 0.

Example Problems

These problems show the answers and solutions.

1. Find the sum of $5 + 6i$ and its conjugate.

   answer: 10

   $5 + 6i$'s conjugate is $5 - 6i$. $(5 + 6i) + (5 - 6i) = (5 + 5) + (6i - 6i) = 10 + 0 = 10$.

2. Find the difference of $5 + 6i$ and its conjugate.

   answer: 12i

   $5 + 6i$'s conjugate is $5 - 6i$. $(5 + 6i) - (5 - 6i) = (5 + 6i) - 5 + 6i = 0 + 12i = 12i$.

3. Find the product of $5 - 6i$ and its conjugate.

   answer: 61

   $5 - 6i$'s conjugate is $5 + 6i$.

   Multiply by the "FOIL" method: $(5 + 6i)(5 - 6i) = 25 + 30i - 30i - 36i^2$.

   Remember that $i^2 = -1$, so $25 + 30i - 30i - 36i^2 = 25 - 36(-1) = 25 + 36 = 61$.

Multiplying and Dividing Complex Numbers

Consider that s and t represent two complex numbers.

$$s = a(\cos\alpha + i\sin\alpha)$$
$$t = b(\cos\beta + i\sin\beta)$$

To multiply s and t, multiply their moduli (plural of modulus) and add their amplitudes.

$$st = ab[\cos(\alpha + \beta) + i\sin(\alpha + \beta)]$$

To find the quotient of two complex numbers, say $s \div t$, divide their moduli and subtract their amplitudes:

$$\frac{s}{t} = \frac{a}{b}\left[\cos(\alpha + \beta) - i\sin(\alpha - \beta)\right]$$

Example Problems

These problems show the answers and solutions. Leave all answers in modulus and amplitude form.

1. If $q = 5(\cos 40° + i\sin 40°)$, and $r = 7(\cos 60° + i\sin 60°)$, then find qr.

   answer: qr = 35(cos100° + isin100°)

   The moduli, 5 and 7, are multiplied together to get 35. The amplitudes (inside the parentheses) are added together, with like terms being added to like terms.

2. If $v = 15(\cos80° + i\sin80°)$, $w = 5(\cos120° + i\sin120°)$. Find $v \div w$.

   answer: 3(cos320° + isin320°)

   First, set up the moduli and amplitudes: $\quad \dfrac{v}{w} = \dfrac{15}{5}[\cos(80° - 120°) + i\sin(80° - 120°)]$

   Next, perform computations: $\quad \dfrac{v}{w} = 3(\cos -40° + i\sin-40°)$

   Rather than leaving negative angles, add 360°: $\quad \dfrac{v}{w} = 3(\cos320° + i\sin320°)$

3. If $l = 4(\cos60° + i\sin60°)$, and $m = 6(\cos75° + i\sin75°)$, then find lm.

   answer: lm = 24(cos135° + isin135°)

   The moduli are multiplied, and the amplitudes added:

   $lm = 4 \cdot 6 [\cos(60° + 75°) + i\sin(60° + 75°)]$

   Finally, perform the computations: $\quad lm = 24(\cos135° + i\sin135°)$

4. If $p = 12(\cos120° + i\sin120°)$, $q = 3(\cos50° + i\sin50°)$. Find $p \div q$.

   answer: 4(cos70° + isin70°)

   First, set up the moduli and amplitudes: $\dfrac{p}{q} = \dfrac{12}{3}[\cos(120° - 50°) + i\sin(120° - 50°)]$

   Next, perform computations: $\dfrac{p}{q} = \dfrac{12}{3}(\cos70° + i\sin70°)$

Finding Powers of Complex Numbers

The formula for finding powers of complex numbers is attributed to a seventeenth century French mathematician named Abraham DeMoivre. Although proving **DeMoivre's theorem** is way beyond the scope of this book, it can be shown to be true for all real-number values of n, and not only the integral values. The formula he developed goes like this:

If a complex number $z = r(\cos\phi + i\sin\phi)$, then $z^2 = r^2(\cos2\phi + i\sin2\phi)$, and $z^3 = r^3(\cos3\phi + i\sin3\phi)$, and so on. The theorem goes on to generalize that for n equal to any positive integer, $z^n = r^n(\cos n\phi + i\sin n\phi)$.

Example Problems

These problems show the answers and solutions.

Use DeMoivre's theorem to solve each of the following.

1. $[\sqrt{2}(\cos40° + i\sin40°)]^5$ Express the result in polar notation.

   answer: $\left(4\sqrt{2}, 200°\right)$

   Apply DeMoivre's theorem: $\quad \left[\sqrt{2}(\cos40° + i\sin40°)\right]^5 = \sqrt{2}^5(\cos40° \cdot 5 + i\sin40° \cdot 5)$

   Multiply: $\quad = \sqrt{2}^5(\cos200° + i\sin200°)$

   Work out $\sqrt{2}^5$: $\quad \sqrt{2} \cdot \sqrt{2} \cdot \sqrt{2} \cdot \sqrt{2} \cdot \sqrt{2} = 2 \cdot 2\sqrt{2} = 4\sqrt{2}$

So: $= 4\sqrt{2}(\cos 200° + i\sin 200°)$

You now know $r = 4\sqrt{2}$ and the angle is in quadrant III. Polar coordinates are $(4\sqrt{2}, 200°)$.

2.   Raise $1 + i$ to the eighth power. Start with a graph.

***answer:* 16**

If you graphed the complex number, you would have noticed that it is an isosceles right triangle, so $\phi = 45°$.

That makes $r = \sqrt{1^2 + 1^2} = \sqrt{2}$.

Apply DeMoivre's theorem: $[\sqrt{2}(\cos 45° + i\sin 45°)]^8 = (\sqrt{2})^8(\cos 8 \cdot 45° + i\sin 8 \cdot 45°)$

Multiply: $= 16(\cos 360° + i\sin 360°)$

$360°$ is the same as $0°$, so find ratios: $= 16(1) + i(16)(0)$

Add to get $16 + 0i$, or 16.

3.   Square $2 + 2i\sqrt{3}$. Express the answer in both rectangular and polar coordinates.

***answer:* (−14, 8*i*); (16, 120°)**

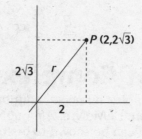

First, find r: $r = \sqrt{2^2 + \left(2\sqrt{3}\right)^2}$

$$r = \sqrt{4 + (4 \cdot 3)} = \sqrt{4 + 12} = \sqrt{16} = 4$$

Now that you've found $r = 4$, use cosine: $\cos\phi = \dfrac{2}{4} = \dfrac{1}{2} = 0.5$.

Find the angle by arccos: $\cos^{-1}\phi = 0.5$; $\phi = 60°$.

Apply DeMoivre's theorem: $[4(\cos 60° + i\sin 60°)]^2 = 4^2(\cos 2 \cdot 60° + i\sin 2 \cdot 60°)$

Multiply: $= 16(\cos 120° + i\sin 120°)$

The reference angle is $30°$, and cos is negative in quadrant II: $= 16(-\cos 30° + i\sin 30°)$

Look up the values and substitute: $= 16(-.866 + 0.5i)$.

Multiply: $= -13.856 + 8i \approx -14 + 8i$.

Work Problems

Use these problems to give yourself additional practice.

1. Find the sum of $9 + 5i$ and its conjugate.

2. Find the difference of $8 + 4i$ and its conjugate.

3. If $l = 5 (\cos 80° + i\sin 80°)$, and $m = 8(\cos 70° + i\sin 70°)$, then find lm.

4. If $v = 14 (\cos 75° + i\sin 75°)$, and $w = 7(\cos 45° + i\sin 45°)$, find $v \div w$.

5. Cube $2 + 2i\sqrt{3}$. Express the answer in both rectangular and polar coordinates.

Worked Solutions

1. **18** $9 + 5i$'s conjugate is $9 - 5i$.

   $(9 + 5i) + (9 - 5i) = (9 + 9) + (5i - 5i) = 18 + 0 = 18$

2. **8i** $8 + 4i$'s conjugate is $8 - 4i$.

   $(8 + 4i) - (8 - 4i) = (8 + 4i) - 8 + 4i = 0 + 8i = 8i$

3. **$lm = 40(\cos 150° + i\sin 150°)$**   The moduli are multiplied, and the amplitudes added:

   $$lm = 5 \cdot 8[\cos(80° + 70°) + i\sin(80° + 70°)]$$

   Finally, perform the computations: $lm = 40(\cos 150° + i\sin 150°)$

4. **$2(\cos 30° + i\sin 30°)$ or $2cis30°$**   First, set up the moduli and amplitudes:

   $$\frac{v}{w} = \frac{14}{7}[\cos(75° - 45°) + i\sin(75° - 45°)]$$

   Next, perform computations: $\frac{v}{w} = 2(\cos 30° + i\sin 30°)$

5. **$(-64, 0)$; $(64, 180°)$**

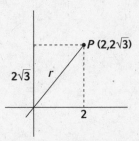

   Graph of Problem 5 at beginning.

   First find r: $r = \sqrt{2^2 + \left(2\sqrt{3}\right)^2}$

   $$r = \sqrt{4 + (4 \cdot 3)} = \sqrt{4 + 12} = \sqrt{16} = 4$$

Now that you've found $r = 4$, use cosine: $\quad \cos\phi = \dfrac{2}{4} = \dfrac{1}{2} = 0.5$

Find the angle by arccos: $\qquad\qquad\qquad\quad \cos^{-1}\phi = 0.5;\ \phi = 60°$

Apply DeMoivre's theorem: $\qquad [4(\cos60° + i\sin60°)]^3 = 4^3\,(\cos3 \cdot 60° + i\sin3 \cdot 60°)$

Multiply: $\qquad\qquad\qquad\qquad\qquad\qquad = 64(\cos180° + i\sin180°)$

The reference angle is 0°, and cos is − in quadrant II: $\quad = 64(-\cos0° + i\sin0°)$

Look up the values and substitute: $\qquad\qquad\quad = 64(-1 + 0i)$

Multiply: $\qquad\qquad\qquad\qquad\qquad\qquad = -64 + 0i.$

In rectangular coordinates, that's (−64, 0); polar is (64, 180°).

Chapter Problems and Solutions

Problems

Solve these problems for more practice applying the skills from this chapter. Worked out solutions follow problems.

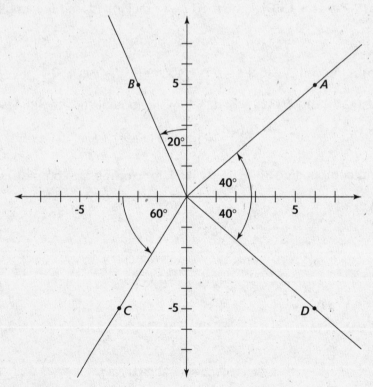

Refer to the preceding figure to answer problems 1 through 4.

1.   What are the polar coordinates of point A?

2.   What are the polar coordinates of point *B*?

3.   What are the polar coordinates of point *C*?

4.   What are the polar coordinates of point *D*?

5.   Convert $P(12, 150°)$ to rectangular coordinates.

6.   Convert $P(6, 45°)$ to rectangular coordinates.

7.   Convert $P(10, 120°)$ to rectangular coordinates.

8.   Convert $P(12, 300°)$ to rectangular coordinates.

9.   Convert $P(8, 8)$ to polar coordinates.

10.   Convert $P(-8, -5)$ to polar coordinates.

11.   Express $P(6, -10)$ in polar coordinates.

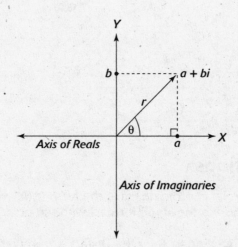

Use the diagram in the preceding figure as a model for plotting the complex numbers in 12–15.

12.   Plot $6 + 4i$ on the complex plane.

13.   Plot $-5 + 4i$ on the complex plane.

14.   Plot $-6 - 2i$ on the polar plane.

15.   Plot $7 - 5i$ on the polar plane.

16.   Find the sum of $8 + 9i$ and its conjugate.

17.   Find the difference of $8 + 7i$ and its conjugate.

18.   Find the product of $6 - 7i$ and its conjugate.

For problems 19–22, leave all answers in polar modulus and amplitude form.

19.   If $s = 5(\cos 75° + i\sin 75°)$ and $t = 3(\cos 45° + i\sin 45°)$, then find st.

20.   If $g = 5(\cos 105° + i\sin 105°)$ and $h = 7(\cos 80° + i\sin 80°)$, then find gh.

21.   If $q = 18(\cos 50° + i\sin 50°)$ and $r = 6(\cos 70° + i\sin 70°)$, then find $q \div r$.

22.   If $m = 15(\cos 150° + i\sin 150°)$ and $n = 3(\cos 60° + i\sin 60°)$, then find $m \div n$.

Use DeMoivre's theorem to solve each of the following.

23. $[\sqrt{3}(\cos 50° + i\sin 50°)]^6$. Express the result in polar notation.

24.   Raise $3 + 3i$ to the fourth power. Write the answer as a complex number.

25.   Cube $5 + 7i$. Express the answer in both complex and polar coordinates.

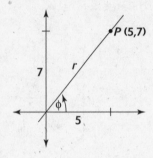

Answers and Solutions

1.   ***Answer:*** $(\sqrt{61}, 40°)$ Polar coordinates take the form (r, θ), where r is the distance from the pole to the point and θ is the angle formed with the polar axis.

Find the distance from the pole to A by the Distance Formula: $r = \sqrt{x^2 + y^2}$

Read values from axes and substitute: $r = \sqrt{6^2 + 5^2}$

Compute: $r = \sqrt{36 + 25} = \sqrt{61}$

θ is read from the graph as 40° counterclockwise, which is positive.

2.   ***Answer:*** $(\sqrt{29}, 110°)$ Polar coordinates take the form (r, θ), where r is the distance from the pole to the point and θ is the angle formed with the polar axis.

Find the distance from the pole to B by the Distance Formula: $r = \sqrt{x^2 + y^2}$

Read values from axes and substitute: $r = \sqrt{(-2)^2 + 5^2}$

Compute: $r = \sqrt{4 + 25} = \sqrt{29}$

θ is read from the graph as 40° + 50° to the vertical, + another 20° for a total θ of 110° counterclockwise, which is positive.

3. ***Answer:*** $(\sqrt{34}, 240°)$　Polar coordinates take the form (r,θ), where r is the distance from the pole to the point and θ is the angle formed with the polar axis.

    Find the distance from the pole to C by the Distance Formula:　　$r = \sqrt{x^2 + y^2}$

    Read values from axes and substitute:　　　　　　　　　　$r = \sqrt{(-3)^2 + (-5)^2}$

    Compute:　　　　　　　　　　　　　　　　　　　　　　$r = \sqrt{9 + 25} = \sqrt{34}$

    θ is computed from the graph as 240° counterclockwise 60° beyond 180°, which is positive.

4. ***Answer:*** $(\sqrt{61}, -40°)$　Polar coordinates take the form (r,θ), where r is the distance from the pole to the point and θ is the angle formed with the polar axis.

    Find the distance from the pole to D by the Distance Formula:　　$r = \sqrt{x^2 + y^2}$

    Read values from axes and substitute:　　　　　　　　　　$r = \sqrt{6^2 + (-5)^2}$

    Compute:　　　　　　　　　　　　　　　　　　　　　　$r = \sqrt{36 + 25} = \sqrt{61}$

    θ is read from the graph as 40° clockwise, which is negative.

5. ***Answer:*** **(−10.4, 6)**

    First, find x using cosine:　　　　　$\frac{x}{r} = \cos\theta$

    Cross multiply:　　　　　　　　　$x = r\cos\theta$

    Substitute (ref∠ = 30°):　　　　　$x = 12\cos\theta°$

    Solve (cos negative in quadrant II):　$x = 12(-0.866) \approx -10.4$

    Similarly for y:　　　　　　　　$\frac{y}{r} = \sin\theta$

    Cross multiply to get:　　　　　　$y = r\sin\theta$

    Substitute:　　　　　　　　　　$y = 12\sin30°$

    Solve (sin is + in quadrant II):　　$y = 12(0.5) = 6$.

    Therefore, $P(12, 150°)$ has rectangular coordinates $P(-10.4, 6)$.

6. ***Answer:*** **(4.2, 4.2)**

    First, find x using cosine:　$\frac{x}{r} = \cos\theta$

    Cross multiply:　　　　　　$x = r\cos\theta$

    Substitute:　　　　　　　$x = 6\cos45°$

    Solve:　　　　　　　　　$x = 6(0.707) = 4.24 \approx 4.2$

    You don't need to go any further, since it's a 45° angle. A 45° right triangle is an isosceles one, so y must also be 4.2.

7. *Answer:* **(–5, 8.7)**

First, find x using cosine: $\qquad\qquad\qquad\qquad \dfrac{x}{r} = \cos\theta$

Cross multiply: $\qquad\qquad\qquad\qquad\qquad\quad x = r\cos\theta$

Substitute (ref. angle = 60°): $\qquad\qquad\quad x = 10\cos60°$

Solve (remembering cos is negative in QII: $x = 10(-0.5) = -5$

Similarly for y: $\qquad\quad \dfrac{y}{r} = \sin\theta$

Cross multiply to get: $y = r\sin\theta$

Substitute: $\qquad\qquad\quad y = 10\sin60°$

Solve: $\qquad\qquad\qquad\quad y = 10(0.866) = 8.66 \approx 8.7$

Therefore, $P(10, 120°)$ has rectangular coordinates $P(-5, 8.7)$.

8. *Answer:* **(6, –10.4)**

First, find x using cosine: $\qquad\qquad\qquad \dfrac{x}{r} = \cos\theta$

Cross multiply: $\qquad\qquad\qquad\qquad\quad x = r\cos\theta$

Substitute (ref$\angle$= 60°): $\qquad\qquad\quad x = 12\cos60°$

Solve (cos positive in quadrant IV): $x = 12(0.5) = 6$

Similarly for y: $\qquad\qquad\qquad \dfrac{y}{r} = \sin\theta$

Cross multiply to get: $\qquad\qquad\quad y = r\sin\theta$

Substitute: $\qquad\qquad\qquad\quad y = 12\sin60°$

Solve (sin is negative in quadrant III): $y = 12(-0.866) = -10.392 \approx -10.4$

Therefore, $P(12, 300°)$ has rectangular coordinates $P(6, -10.4)$.

9. *Answer:* $(8\sqrt{2}, 45°)$

First, use the Pythagorean theorem: $r^2 = x^2 + y^2$

Solve that for r: $\qquad\qquad\qquad r = \sqrt{x^2 + y^2}$

Substitute: $\qquad\qquad\qquad\qquad r = \sqrt{8^2 + 8^2}$

Square terms indicated: $\qquad\qquad r = \sqrt{64 + 64}$

Add and take the square root: $\qquad r = \sqrt{128} = 8\sqrt{2}$

Don't bother with a trigonometric ratio, since the x and y coordinates were the same. The angle is 45°. Think about it: isosceles right triangle; legs equal.

The polar coordinates for $P(8,8)$: $(r, \theta) = (8\sqrt{2}, 45°)$

10. **Answer:** $(\sqrt{89}, 212°)$

Start with the formula: $r^2 = x^2 + y^2$

Solve that for r: $r = \sqrt{x^2 + y^2}$

Substitute: $r = \sqrt{(-8)^2 + (-5)^2}$

Square terms indicated: $r = \sqrt{64 + 25}$

Add and try to simplify: $r = \sqrt{89}$

Next, use the tangent ratio: $\tan\theta = \dfrac{y}{x}$

Substitute: $\tan\theta = \dfrac{-5}{-8} = 0.625$

Use $\tan^{-1}\theta$ to find the angle: $\tan^{-1}\theta = 0.625; \theta \approx 32°$

But the coordinates placed the angle in quadrant III, so add 180° to the angle you just found.

$180° + 32° = 212°$ so $P(-8, -5)$ in polar coordinates: $(r, \theta) = (\sqrt{89}, 212°)$

11. **Answer:** $(2\sqrt{34}, 301°)$

First, use the formula: $r^2 = x^2 + y^2$

Solve that for r: $r = \sqrt{x^2 + y^2}$

Substitute: $r = \sqrt{6^2 + (-10)^2}$

Square terms indicated: $r = \sqrt{36 + 100}$

Add and simplify: $r = \sqrt{136} = 2\sqrt{34}$

Next, pick the tangent ratio: $\tan\theta = \dfrac{y}{x}$

Substitute: $\tan\theta = \dfrac{-10}{6} = -1.667$

Use $\tan^{-1}\theta$ to find the first quadrant I angle: $\tan^{-1}\theta = 1.667; \theta \approx 59°$

But $P(6, -10)$ is in quadrant IV, which is why tan is negative. Reference angle is $-59°$, or 301°.

The polar coordinates for $P(6, -10)$ are: $(r, \theta) = (2\sqrt{34}, 301°)$

12.   ***Answer: A***   **See the following figure.**

Both signs are positive, so it goes in quadrant I.

13.   ***Answer: B***   **See the following figure.**

A negative *a* coordinate and a positive *b* coordinate put the number in the second quadrant.

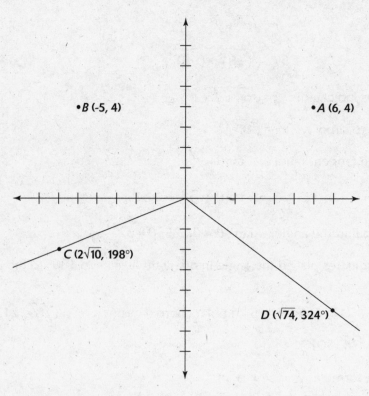

14.   ***Answer: C***   **See the preceding figure.** The −6 is the horizontal coordinate and −2*i* is in the vertical coordinate. To find *r*, you'll use the Distance Formula version of the Pythagorean theorem.

Formula: $r = \sqrt{x^2 + y^2}$

Substitute: $r = \sqrt{(-6)^2 + (-2)^2}$

Square: $r = \sqrt{36 + 4}$

Simplify: $r = \sqrt{40} = \sqrt{4 \cdot 10} = \sqrt{4} \cdot \sqrt{10} = 2\sqrt{10}$

Use tangent to find the angle: $\tan\theta = \dfrac{2}{6} = 0.333$

Find the angle with table or calculator: $\tan^{-1}\theta = 0.333$; $\theta = 18.42°$

The angle is about 18°, but, if you look back at the rectangular coordinates, it is in the third quadrant. That means you must add another 180° to the 18 to get 198°.

15. ***Answer: D   See the preceding figure.*** The 7 is the horizontal coordinate, and −5i is the vertical coordinate. To find r, use the Distance Formula version of the Pythagorean theorem.

    Formula: $r = \sqrt{x^2 + y^2}$

    Substitute: $r = \sqrt{7^2 + (-5)^2}$

    Square: $r = \sqrt{49 + 25}$

    Simplify: $r = \sqrt{74}$

    Use tangent to find the angle: $\tan\theta = \dfrac{-5}{7} = -0.714$

    Use tan's absolute value to find the angle: $\tan^{-1}\theta = 0.714$; $\theta = 35.53°$

    The angle is about 36°, but, if you look back at the rectangular coordinates, it is in the fourth quadrant. That means you must subtract it from 360°. 360° − 36° = 324°.

16. ***Answer: 16***   8 + 9i's conjugate is 8 − 9i.

    (8 + 9i) + (8 − 9i) = (8 + 8) + (9i − 9i) = 16 + 0 = 16

17. ***Answer: 14i***   8 + 7i's conjugate is 8 − 7i.

    (8 + 7i) − (8 − 7i) = (8 + 7i) − 8 + 7i = 0 + 14i = 14i

18. ***Answer: 85***   6 − 7i's conjugate is 6 + 7i.

    Multiply by the "FOIL" method: $(6 + 7i)(6 - 7i) = 36 + 42i - 42i - 49i^2$

    Remember that $i^2 = -1$ $i^2 = -1$, so: 36 + 42i − 42i − 49(−1) = 36 + 0 + 49 = 85

19. ***Answer:*** *st* = 15(cos120° + *i*sin120°).

    The moduli are multiplied, and the amplitudes added:

    $$st = 5 \cdot 3\,[(\cos 75° + 45°) + i\sin(75° + 45°)]$$

    Finally, perform the computations: *st* = 15(cos120° + *i*sin120°)

20. ***Answer:*** *gh* = 35(cos185° + *i*sin185°).

    The moduli are multiplied, and the amplitudes added:

    $$gh = 5 \cdot 7\,[(\cos 105° + 80°) + i\sin(105° + 80°)]$$

    Finally, perform the computations: *gh* = 35(cos185° + *i*sin185°)

21. ***Answer:*** 3(cos340° + *i*sin 340°).

    First, set up the moduli and amplitudes: $\dfrac{q}{r} = \dfrac{18}{6}[\cos(50° - 70°) + i\sin(50° - 70°)]$

    Next, perform computations: $\dfrac{q}{r} = 3(\cos{-20°} + i\sin{-20°})$

    Rather than leaving negative angles, add 360°: $\dfrac{q}{r} = 3(\cos 340° + i\sin 340°)$

22. ***Answer:*** $5(\cos 90° + i\sin 90°)$.

First, set up the moduli and amplitudes: $\frac{m}{n} = \frac{15}{3}[\cos(150° - 60°) + i\sin(150° - 60°)]$

Next, perform computations: $\frac{m}{n} = 5(\cos 90° + i\sin 90°)$

23. ***Answer:*** $(27, 300°)$.

Apply DeMoivre's theorem: $[\sqrt{3}(\cos 50° + i\sin 50°)]^6 = (\sqrt{3})^6(\cos 6 \cdot 50° + i\sin 6 \cdot 50°)$

Multiply: $(\sqrt{3})^6(\cos 300° + i\sin 300°)$

Work out $(\sqrt{3})^6$: $\sqrt{3} \cdot \sqrt{3} \cdot \sqrt{3} \cdot \sqrt{3} \cdot \sqrt{3} \cdot \sqrt{3} = 3 \cdot 3 \cdot 3 = 27$

The result is: $= 27(\cos 300° + i\sin 300°)$

You now know $r = 27$ and the angle is in quadrant IV. Polar coordinates are $(27, 300°)$

24. ***Answer:*** -324 or $-324 + 0i$

If you graphed the complex number, you would have noticed that it is an isosceles right triangle, so $\phi = 45°$. That makes $r = \sqrt{3^2 + 3^2} = \sqrt{18} = 3\sqrt{2}$.

Apply DeMoivre's theorem: $[3\sqrt{2}(\cos 45° + i\sin 45°)]^4 = (3\sqrt{2})^4(\cos 4 \cdot 45° + i\sin 4 \cdot 45°)$

Multiply: $= (3\sqrt{2})^4(\cos 180° + i\sin 180°)$

Work out $(3\sqrt{2})^4$: $(3\sqrt{2})^4 = 3^4(\sqrt{2}) = 81 \cdot 4 = 324$

The result is: $= 324(\cos 180° + i\sin 180°)$

Reference angle for 180° is 0°, with cos negative: $= 324(-1) + i(324)(0)$

Add to get $-324 + 0i$, or -324.

25. ***Answer:*** $(-610 + 181i)$; $(636, 163.5°)$.

First, find r: $r = \sqrt{5^2 + 7^2}$

$$r = \sqrt{25 + 49} = \sqrt{74} \approx 8.6$$

Since r isn't exact, use tan to find ϕ: $\tan\phi = \frac{7}{5} = 1.4$

Find the angle by arctan: $\tan^{-1}\phi = 1.4$; $\phi \approx 54.5°$

Apply DeMoivre's theorem:

$$[8.6(\cos 54.5° + i\sin 54.5°)]^3 = 8.6^3(\cos 3 \cdot 54.5° + i\sin 3 \cdot 54.5°)$$

Multiply: $= 636(\cos 163.5° + i\sin 163.5°)$

The reference angle is 16.5°, and cos is negative in quadrant II:

$$= 636(-\cos 16.5° + i\sin 16.5°)$$

Look up the values and substitute: $= 636(-0.959 + 0.284i)$

Multiply: $\approx -610 + 181i$.

Supplemental Chapter Problems

Solve these problems for even more practice applying the skills from this chapter. The Answer section will direct you to where you need to review.

Problems

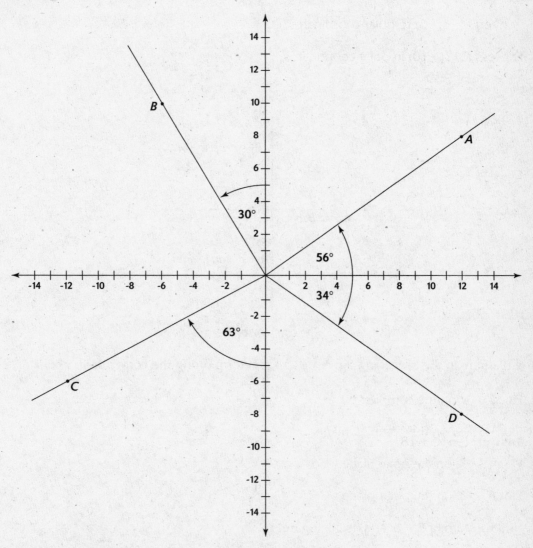

Refer to the preceding figure to answer problems 1 through 4.

1. What are the polar coordinates of point *A*?

2. What are the polar coordinates of point *B*?

3. What are the polar coordinates of point *C*?

4. What are the polar coordinates of point *D*?

5. Convert $P(18, 60°)$ to rectangular coordinates.

6. Convert $P(12, 45°)$ to rectangular coordinates.

7. Convert $P(8, 120°)$ to rectangular coordinates.

8. Convert $P(16, 210°)$ to rectangular coordinates.

9. Convert $P(12, 12)$ to polar coordinates.

10. Convert $P(-8, -6)$ to polar coordinates.

11. Express $P(12, -10)$ in polar coordinates.

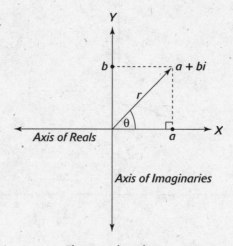

The complex plane.

Use the diagram in the preceding figure as a model for plotting the complex numbers in 12–15.

12. Plot $7 + 5i$ on the complex plane.

13. Plot $-6 + 9i$ on the complex plane.

14. Plot $-8 - 3i$ on the polar plane.

15. Plot $6 - 10i$ on the polar plane.

16. Find the sum of $5 + 6i$ and its conjugate.

17. Find the difference of $9 + 8i$ and its conjugate.

18. Find the product of $5 - 4i$ and its conjugate.

For problems 19–22, write all answers in *cis* form.

19. If $s = 7(\cos 55° + i\sin 55)$ and $t = 4(\cos 65° + i\sin 65°)$, then find st.

20. If $g = 3(\cos 95° + i\sin 95)$ and $h = 5(\cos 65° + i\sin 65°)$, then find gh.

21. If $q = 16(\cos 50° + i\sin 50)$ and $r = 4(\cos 100° + i\sin 100°)$, then find $q \div r$.

22. If $m = 40(\cos120° + i\sin120)$ and $n = 8(\cos70° + i\sin 70°)$, then find $m \div n$.

Use DeMoivre's theorem to solve each of the following.

23. $[3(\cos70° + i\sin70)]^4$. Express the result in polar notation.

24. Raise $3\sqrt{3} + 3i$ to the third power. Write the answer as a complex number.

25. Raise $3 + 5i$ to the 5th power. Express the answer in both rectangular and polar coordinates.

Answers

1. $(\sqrt{208}, 56°)$ (Polar Coordinates, p. 203)

2. $(2\sqrt{34}, 120°)$ (Polar Coordinates, p. 203)

3. $(6\sqrt{5}, 207°)$ (Polar Coordinates, p. 203)

4. $(4\sqrt{13}, 326°)$ (Polar Coordinates, p. 203)

5. $(9, 15.6)$ (Converting from Polar to Rectangular Coordinates, p. 206)

6. $(8.5, 8.5)$ (Converting from Polar to Rectangular Coordinates, p. 206)

7. $(-4, 6.9)$ (Converting from Polar to Rectangular Coordinates, p. 206)

8. $(-13.9, -8)$ (Converting from Polar to Rectangular Coordinates, p. 206)

9. $(12\sqrt{2}, 45°)$ (Converting from Rectangular to Polar Coordinates, p. 206)

10. $(10, 216.9°)$ (Converting from Rectangular to Polar Coordinates, p. 206)

11. $(2\sqrt{61}, 320.2°)$ (Converting from Rectangular to Polar Coordinates, p. 206)

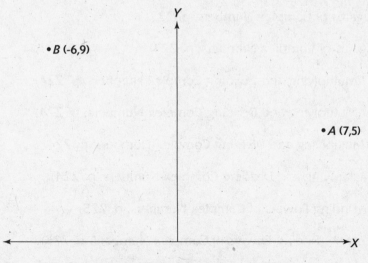

Answers for Problems 12 and 13.

12. See the preceding figure. (Plotting Complex Numbers on Rectangular Axes, p. 217)

13. See the preceding figure. (Plotting Complex Numbers on Rectangular Axes, p. 217)

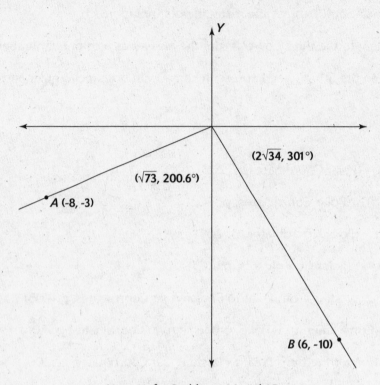

Answers for Problems 14 and 15.

14. *A* (See the preceding figure). (Plotting Complex Numbers on the Polar Axis, p. 219)

15. *B* (See the preceding figure). (Plotting Complex Numbers on the Polar Axis, p. 219)

16. 10 (Conjugates of Complex Numbers, p. 223)

17. 16*i* (Conjugates of Complex Numbers, p. 223)

18. 41 (Conjugates of Complex Numbers, p. 223)

19. 28*cis*120° (Multiplying and Dividing Complex Numbers, p. 224)

20. 15*cis*160° (Multiplying and Dividing Complex Numbers, p. 224)

21. 4*cis*310° (Multiplying and Dividing Complex Numbers, p. 224)

22. 5*cis*50° (Multiplying and Dividing Complex Numbers, p. 224)

23. (81, 280°) (Finding Powers of Complex Numbers, p. 225)

24. 0 + 216*i* or 216*i* (Finding Powers of Complex Numbers, p. 225)

25. (2849, −6109); (6741, 295°) (Finding Powers of Complex Numbers, p. 225)

Chapter 7

Inverse Functions and Equations

We discussed much earlier in this book the six basic trigonometric functions. In order for a mathematic relation to be a function, there cannot be more than one y-value for any x-value. Consider the graph of the function $f(x) = x^2$. You should remember it from Algebra I. It's an upward opening parabola with its vertex at the origin, as shown in the following figure.

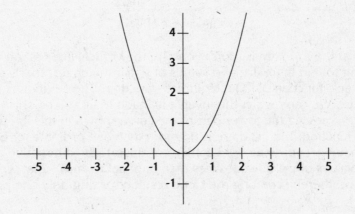

$y = x^2$ A typical Algebra I parabola.

Its equation is usually written $y = x^2$. It is a true function since it passes the "vertical line test". In other words, any vertical line drawn to intersect the graph will encounter one and only one point on the graph.

Restricting Functions

The notation $f^{-1}(x)$ is used to designate the **inverse** of a function. Don't confuse the -1 with the exponent -1. If $f(x) = x^2$, $f^{-1}(x)$ **is not** $\dfrac{1}{x^2}$. Rather, $f^{-1}(x)$ is the inverse, $x = y^2$, as seen in the figure that follows.

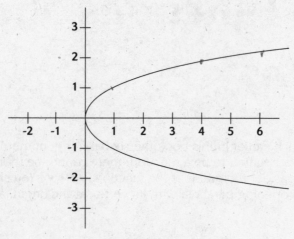

$x = y^2$ The inverse of $y = x^2$.

But now, you see what the problem is, don't you? Try the vertical line test on any positive value of x, and you're going to find there are two values of y that match up. That's not allowed. How can we make the inverse function of $f(x)$ a legitimate function? The solution is to **restrict** it by removing half the values. We can restrict the inverse function to all values of y that are equal to or greater than zero. By removing the lower part of the curve shown in this figure, we'll create an inverse that is a true function. That is, there is no more than one ordinate for each abcissa (that means no more than one y coordinate for each x coordinate). The alternative is to restrict the inverse function to all values of $y \le 0$ (values less than or equal to zero). That would remove the upper half of the curve, thereby creating the same result of pairing only one ordinate with one abcissa.

Inverses of trigonometric functions are relations that are not functions at all, because there is more than one distinct value of y for each value of x. Consider the inverses of the sine and the cosine in the following figure.

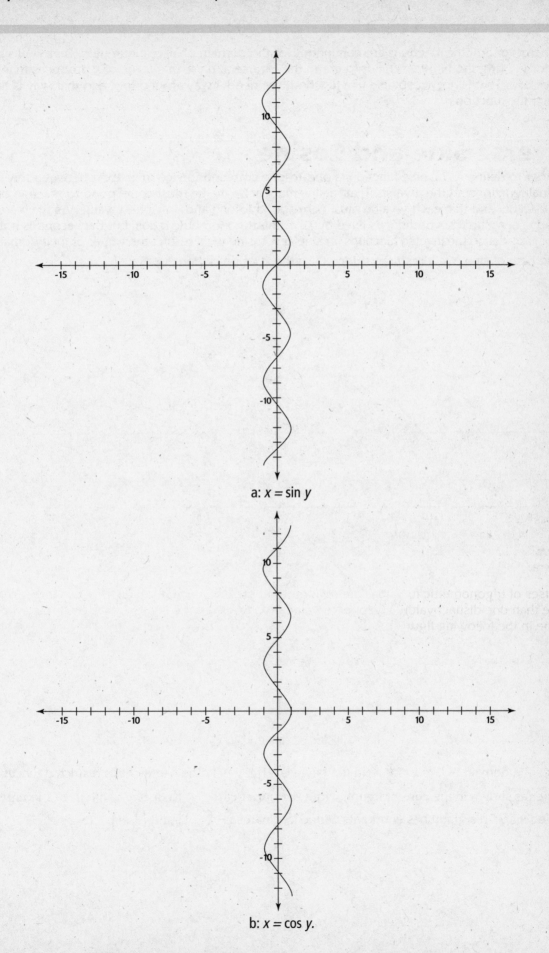

a: $x = \sin y$

b: $x = \cos y$.

All the trigonometric functions are continuous for the domain of all real numbers. That works out fine for $y = $ (trigonometric) $f(x)$, but to graph the inverse and the one value of x produces many values of y. That is not acceptable in a function. We need a way around this, and that way is to restrict the function.

Inverse Sine and Cosine

In order to define an inverse function, a one-to-one correspondence must exist between the original function and the inverse. That means that each x-value must correspond to one and only one y-value, and that each y-value must correspond to one and only one x-value. As we have already seen, the first condition is filled by all of the trigonometric ratios, but the second is not. In order to define the inverse functions for sine and cosine we'll restrict the values of their domains. In the case of sine, we'll restrict its domain to those values where $-\frac{\pi}{2} \le x \le \frac{\pi}{2}$.

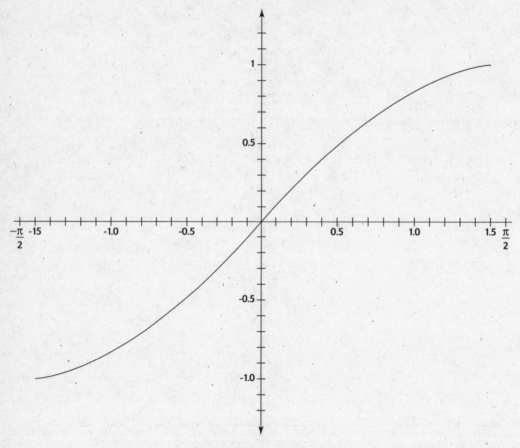

Restricted sine $y = \mathrm{Sin}\ x$.

Notice the domain: $-\frac{\pi}{2} \le x \le \frac{\pi}{2}$, and the range is $-1 \le y \le 1$. The name of this restricted function is Sine (as shown in the current figure). Note the upper case "S" to distinguish it from the unrestricted sine. This graph has endpoints with coordinates $\left(-\frac{\pi}{2}, -1\right)$ and $\left(\frac{\pi}{2}, 1\right)$.

The **inverse sine function** can now be defined in terms of the inverse of the restricted Sine function, $y = \text{Sin } x$ (see the following figure).

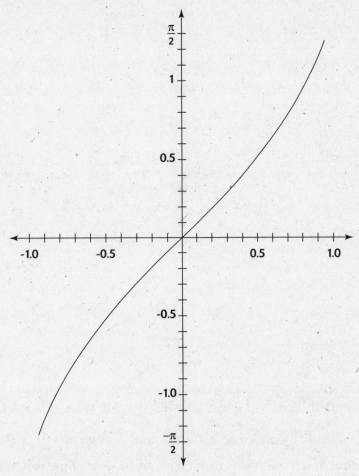

Graph of Inverse of Restricted Sine: $y = \text{Sin}^{-1} x$.

Note that for inverse sine, the domain is restricted to $-1 \leq x \leq 1$ and the range is $-\frac{\pi}{2} \leq y \leq \frac{\pi}{2}$. Compare those to the domain and range of $y = \text{Sin } x$, above.

The identities for the sine and inverse sine are:

$$\sin(\text{Sin}^{-1}x) = x \qquad -1 \leq x \leq 1$$

$$\text{Sin}^{-1}(\sin x) = x \qquad -\frac{\pi}{2} \leq x \leq \frac{\pi}{2}$$

Now let's look at restricted cosine. Check out the figure that follows.

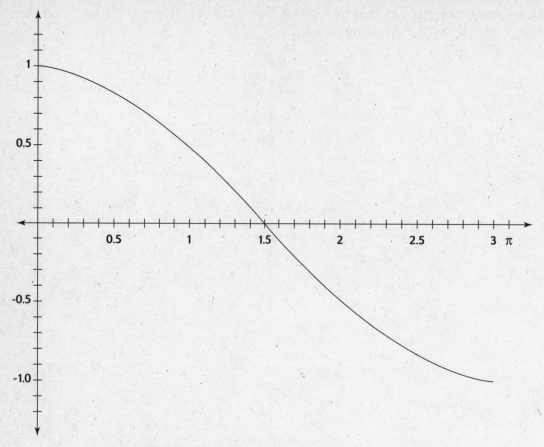

Restricted Cosine from 0 to π; Domain $0 \leq x \leq \pi$; range $-1 \leq y \leq 1$.

The restrictions on cosine are a domain of $0 \leq x \leq \pi$, and a range of $-1 \leq y \leq 1$.

The **inverse cosine function** is, naturally, defined as the inverse of the restricted Cosine function $\text{Cos}^{-1}(\cos x) = x$ where $0 \leq x \leq \pi$. It follows then that:

$$y = \text{Cos}^{-1}x, \text{ when } -1 \leq x \leq 1 \text{ and } 0 \leq y \leq \pi,$$

as shown in the following figure. Once again, notice that the x and y values are exchanged in the inverse from what they were in the original function.

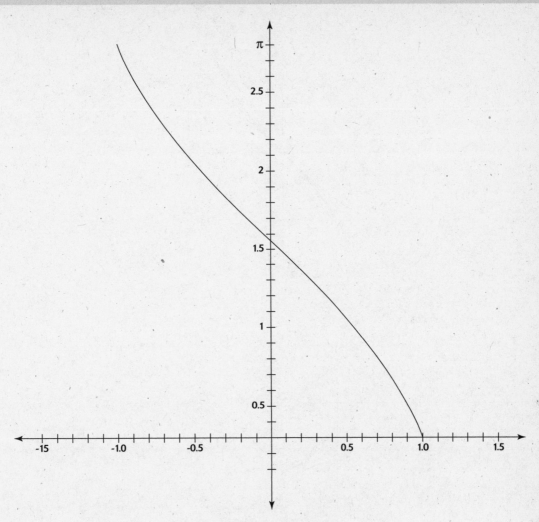

Restricted inverse cosine.

The domain is $-1 \leq x \leq 1$, and the range is $0 \leq y \leq \pi$, as already noted.

The identities for the cosine and inverse cosine are:

$$\cos(\text{Cos}^{-1}x) = x \qquad -1 \leq x \leq 1$$

$$\text{Cos}^{-1}(\cos x) = x \qquad -\frac{\pi}{2} \leq x \leq \frac{\pi}{2}$$

If we were to graph $y = \text{Sin}\,x$ and $y = \text{Sin}^{-1}x$ on a single set of axes, we'd find the two curves are symmetric over the line $x = y$. The same is true for $y = \text{Cos}\,x$ and $y = \text{Cos}^{-1}x$, as you can see in the figure that follows.

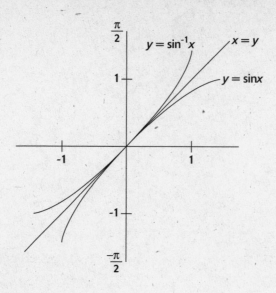

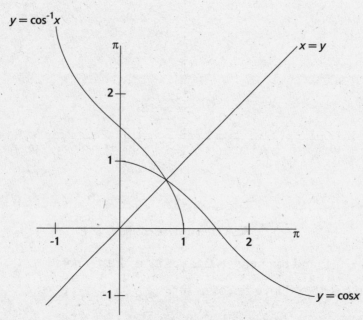

Symmetry of inverses.

An inverse trig function behaves like the inverse of any other type of function, which is to undo what the original function did. Sin⁻¹x may also be written Arcsinx, and Cos⁻¹x may be written Arccosx.

Example Problems

These problems show the answers and solutions.

1. Find the exact value of $\text{Cos}\left(-\dfrac{\sqrt{2}}{2}\right)$

    answer: 135° or $\dfrac{3\pi}{4}$

    The rationale one uses to solve trigonometric inverses is not the same as one would use on an algebraic equation.

    Suppose $y = \text{Cos}\left(-\dfrac{\sqrt{2}}{2}\right)$. That means that $\cos y = -\dfrac{\sqrt{2}}{2}$ where $0 \le y \le \pi$.

    We should recognize $\dfrac{\sqrt{2}}{2}$ to be the cosine of 45°, or $\dfrac{\pi}{4}$ radians. Since the cosine is $-\dfrac{\sqrt{2}}{2}$, it must be in the second quadrant. Subtract 180° − 45° = 135°. In radians, $\pi - \dfrac{\pi}{4} = \dfrac{3\pi}{4}$.

2. Find the exact value of $\text{Sin}^{-1}\left(\dfrac{1}{2}\right)$

    answer: 30° or $\dfrac{\pi}{6}$

    Suppose $y = \text{Sin}^{-1}\left(\dfrac{1}{2}\right)$. That means that $\sin y = y = \dfrac{1}{2}$ where $-1 \le y \le 1$. $\dfrac{1}{2}$ is the sin of 30°, or $\dfrac{\pi}{6}$.

3. Find the exact value of $\cos(\text{Cos}^{-1} 0.73)$.

    answer: 0.73

    Use the cosine-inverse cosine identity: $\cos(\text{Cos}^{-1} x) = x$ $-1 \le x \le 1$

    Therefore: $\cos(\text{Cos}^{-1} 0.73) = 0.73$

Inverse Tangent

Before we can define the **inverse tangent function,** we must first restrict the tangent's domain to those values greater than −90° and less than 90°. That's $-\dfrac{\pi}{2} < x < \dfrac{\pi}{2}$. Take note of the fact that neither opening nor closing values are included in the domain. If you recall the asymptotes that are part of the graph of tangent you'll understand why. This new, restricted tangent is known as Tangent, or, simply, Tan (see the figure that follows).

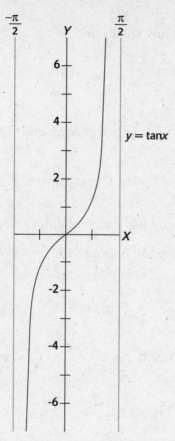

Restricted tangent function.

The inverse tangent function is derived from the restricted Tangent function, $y = \text{Tan } x$.

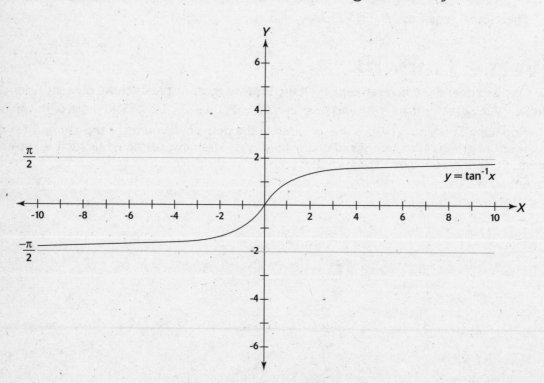

The inverse tangent function.

Since $y = \text{Tan}x$ had the restriction $-\frac{\pi}{2} < x < \frac{\pi}{2}$, $y = \text{Tan}^{-1}x$, has restriction $-\frac{\pi}{2} < y < \frac{\pi}{2}$.

The domain of Arctan is the realm of real numbers, $-\infty < x < \infty$, while the range is $-\frac{\pi}{2} < y < \frac{\pi}{2}$, with the two end values omitted. Remember, tangent is not defined for $90°$ or $-90°$.

The identities for inverse tangent are as follow:

$$\tan(\text{Tan}^{-1}x) = x \qquad -\infty < x < \infty$$

$$\text{Tan}^{-1}(\tan x) = x \qquad -\frac{\pi}{2} \le x \le \frac{\pi}{2}$$

Example Problems
These problems show the answers and solutions.

1. Find the exact value of $\tan\left(\text{Tan}^{-1}\dfrac{1}{\sqrt{3}}\right)$.

    answer: $30°$ or $\dfrac{\pi}{6}$

    No matter whether it's sine, cosine, or tangent, an inverse undoes a function. Regardless of the fact that they appear here in standard form, there is still a function and its inverse, so the answer is whatever $\dfrac{1}{\sqrt{3}}$ is the tangent of. If you recall the 30-60-90 triangle, the side opposite the $30°$ angle is half the hypotenuse, and the side opposite the $60°$ angle is half the hypotenuse times $\sqrt{3}$. See the following figure.

    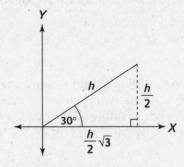

    The 30-60-90 triangle.

    That makes the tangent of $30°$ $\dfrac{\frac{h}{2}}{\frac{h}{2}\sqrt{3}} = \dfrac{1}{\sqrt{3}}$, hence the answer is $30°$, or, in radians, $\dfrac{\pi}{6}$.

2. Find the exact value of $\text{Tan}^{-1}(1)$.

    answer: $45°$ or $\dfrac{\pi}{4}$

    The only way to get a tan = 1 is to have both the opposite and adjacent sides of the right triangle the same size. That makes it an isosceles right triangle. In turn, by definition, both non-right angles of an isosceles right triangle have degree measures of $45°$.

3. Find the exact value of $\text{Tan}^{-1}\left(-\sqrt{3}\right)$.

answer: 300° or $\frac{11\pi}{6}$

To solve this one, refer back to the figure, but imagine that the acute angles have been switched, so that the 60° angle is where the 30° one is in that figure. The side opposite the 60° angle is still half the hypotenuse times $\sqrt{3}$, but this time it's on top: $\dfrac{\frac{h}{2}\sqrt{3}}{\frac{h}{2}} = \sqrt{3}$, so the angle is 60°. But wait. The tangent was $-\sqrt{3}$. The range of Tan^{-1}, you'll recall, is $-\frac{\pi}{2} < y < \frac{\pi}{2}$. That means the first and fourth quadrants. Tangent is negative in Q-IV, so 60° is the reference angle, but we will not leave it as −60°, but will add one rotation and make it −60° + 360° = 300°.

Work Problems

Use these problems to give yourself additional practice.

1. Why is it necessary to restrict a trigonometric function in order to find its inverse function?

2. What is the effect of an inverse function on the original function?

3. Evaluate $\tan(\arcsin\frac{3}{5})$ without using a calculator or table of functions.

4. Write the first six angles to satisfy the condition, $\cos^{-1}\left(\frac{1}{2}\right)$.

5. Find the value of $\text{Tan}^{-1}(0.5)$ to the nearest tenth of a degree.

Worked Solutions

1. **A function is defined as having only a single y value for every x value.** Since trigonometric functions repeat as we go around and around the unit circle, they produce many x values for each y value. **That is permissible.** But once the function is turned around to get the inverse, many y values would be produced for each x value — a situation that is **not** allowed.

2. **It undoes the original function.**

3. $\frac{3}{4}$ If you draw a right triangle, you'll find that it is a triangle whose sine is $\frac{3}{5}$. That means its opposite side is 3 and its hypotenuse is 5. You might recognize that as the basic 3-4-5 right triangle.

4. **60-, 300-, 420-, 660-, 780-, 1020-**   Cosine is positive in Q-I and Q-IV. So the primary values of $\cos\left(\frac{1}{2}\right)$ (those encountered in the first rotation) are 60° and 300°. Then add 360° to each of those for values 3 and 4 (420° and 660°), and another 360° for values 5 and 6 (780° and 1020°).

5. **26.6°**   The question asks for the angle whose tangent is 0.5. Using a calculator or the table (the latter of which requires some interpolation), we find it is 26.6°.

Inverses of Reciprocal Functions

The inverse cosecant, inverse secant, and inverse tangent functions are found by utilizing the restricted Sine, Cosine, and Tangent functions.

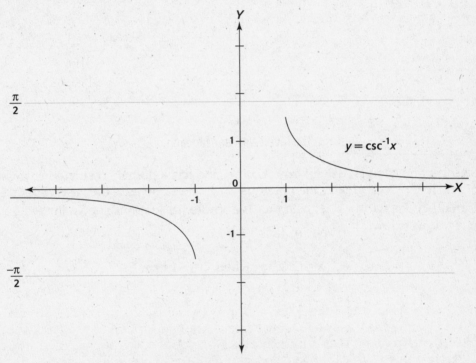

The inverse cosecant function.

The inverse cosecant function is derived from Sine, and has a domain restricted to x-values equal to or less than -1, or greater than or equal to 1. The range is from $y = -\frac{\pi}{2}$ to $\frac{\pi}{2}$ with $y \neq 0$, since that's its asymptote. Its graph is drawn in the preceding figure on page 251 (The 30-60-90 Triangle).

Its identity is as follows:

$$\text{Csc}^{-1}x = \text{Sin}^{-1}\frac{1}{x} \quad \text{when } x \leq -1 \text{ or } x \geq 1 \text{ and } y \neq 0$$

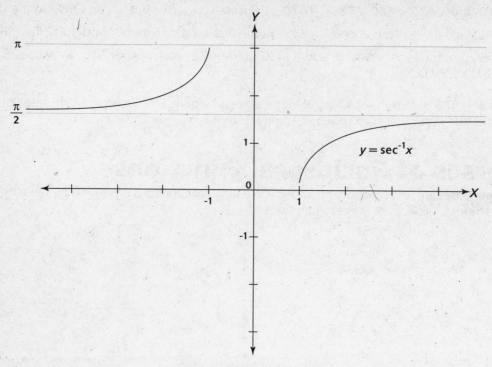

The inverse secant function.

The inverse secant function is derived from Cosine, and has a domain restricted to x-values equal to or less than -1, or greater than or equal to 1, the same as is the case with Csc^{-1}. The range is from $0 \leq y \leq \pi$ with $y \neq \frac{\pi}{2}$, being its asymptote. Its graph is shown in the above figure and its identity is as follows:

$$\text{Sec}^{-1}x = \text{Cos}^{-1}\frac{1}{x} \quad \text{when } x \leq -1 \text{ or } x \geq 1$$

Finally, we get to the inverse cotangent:

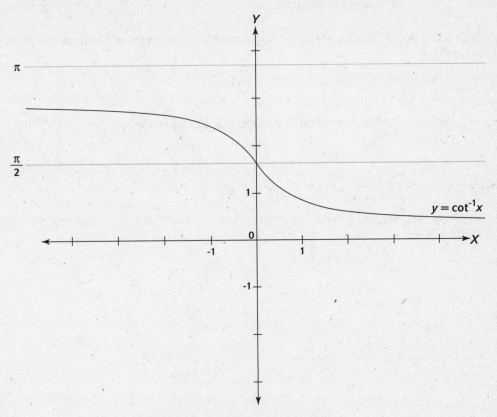

The inverse cotangent function.

The inverse cotangent function is derived from Tangent, and has a domain (*x*-values) equal to all real numbers. The range is from $0 \leq y \leq \pi$ with $y \neq \pi$ and $y \neq 0$, those horizontal lines being its two vertical asymptotes. Its graph is shown in the above figure.

Two identities apply, as follows:

$$\text{Cot}^{-1}x = \text{Tan}^{-1}\frac{1}{x} \quad \text{when } x > 0$$

$$\text{Cot}^{-1}x = \pi + \text{Tan}^{-1}\frac{1}{x} \quad \text{when } x < 0$$

Example Problems

These problems show the answers and solutions.

1. Without using a calculator or table of trigonometric functions, determine the precise value of sin $\operatorname{Sec}^{-1}(-6)$.

   answer: $\dfrac{\sqrt{35}}{6}$

   Suppose we set the bracketed expression equal to the angle, ϕ.

   Then: $\qquad\qquad \phi = \operatorname{Sec}^{-1}(-6)$

   That means: $\quad \operatorname{Sec}\phi = -6$, where $0 \le \phi \le \pi$, $\phi \ne \dfrac{\pi}{2}$

   In this range, secant and cosine are both negative in Q-II. It's time to draw this; see the following figure.

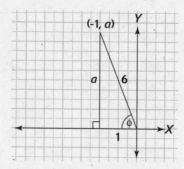

   What we know so far in Problem 1.

   From the reference triangle in the figure, we can use the Pythagorean Theorem to find the third side:

   State the theorem: $\qquad a^2 + b^2 = c^2$

   Substitute: $\qquad\qquad a^2 + (-1)^2 = 6^2$

   Square: $\qquad\qquad\quad a^2 + 1 = 36$

   Collect terms: $\qquad\quad a^2 = 36 - 1$

   Subtract and take the root: $\quad a = \sqrt{35}$

   Therefore: $\quad \sin[\operatorname{Sec}^{-1}(-6)] = \sin\phi$

   And: $\quad \sin[\operatorname{Sec}^{-1}(-6)] = \dfrac{a}{6}$

   Finally: $\quad = \dfrac{\sqrt{35}}{6}$

   And, as a footnote, sine is positive in the second quadrant.

2. Without using a calculator or table of trigonometric functions, determine the precise value of $\cos(\text{Tan}^{-1}8)$.

    answer: $\dfrac{\sqrt{65}}{65}$

    Suppose we set the parenthesized expression equal to the angle ϕ.

    Then: $\phi = \text{Tan}^{-1}8$

    That means: $\text{Tan}\,\phi = 8$, where $-\dfrac{\pi}{2} < \phi < \dfrac{\pi}{2}$

    In this range, tangent and cotangent are both positive in Q-I. It's time to draw this; see the figure that follows.

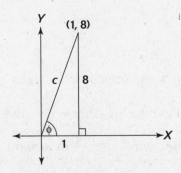

    What we know so far in Problem 2.

    From the reference triangle in this figure, we can use the Pythagorean Theorem to find the third side:

    State the theorem: $c^2 = a^2 + b^2$

    Substitute: $c^2 = 8^2 + 1^2$

    Square: $c^2 = 64 + 1$

    Add and take the root: $c = \sqrt{65}$

    Therefore: $\cos(\text{Tan}^{-1}8) = \cos\phi$

    And: $\cos(\text{Tan}^{-1}8) = \dfrac{1}{\sqrt{65}}$

    Rationalize the denominator: $= \dfrac{1}{\sqrt{65}} \cdot \dfrac{\sqrt{65}}{\sqrt{65}} = \dfrac{\sqrt{65}}{65}$

Trigonometric Equations

Trigonometric equations are strange ducks. Trigonometric identities are true for any angle in the domain of the function or replacement value for the variable involved. Trigonometric equations, also known as **conditional trigonometric equations** are true for only some angles or replacement values, if they are true for any at all. The solutions in a specific interval, usually 0° to 360°, are called the **primary solutions.** A formula that names all possible solutions is called the **general solution.**

There's no clear-cut sequence of steps for solving a trigonometric equation, unlike the ones normally followed for an algebraic one. In an algebraic equation we try to group variables on one side of the equal sign and constants on the other (known as **collecting terms**). Then we perform any operations in parentheses, add, subtract, and finally multiply or divide to solve for the variable, or in the case of a quadratic equation, factor or use the quadratic formula. Some of those might work sometimes in a trigonometric equation, but—even more often—they will not. There are some general guidelines we might follow, that often may help lead to a solution. Successfully solving most trig equations involves plugging in some trigonometric identities at the proper time. Some factoring may also come in handy.

If more than one trigonometric function is involved in an equation, try using identities and algebraic manipulation to get the equation in terms of a single trig function. Look for expressions that can be arranged in quadratic form and then factored. Keep in mind that not all equations have solutions. Try to recognize patterns, and remember that the more experience you have, the easier it will become.

Example Problems

These problems show the answers and solutions.

1. Find the exact values of θ in the equation $\sin^2\theta - \cos\theta = \cos^2\theta$ for $0 \le \theta \le 2\pi$.

   answer: $\theta = 60°, 180°, 300°$, or $\dfrac{\pi}{3}, \pi, \dfrac{5\pi}{3}$ **in radians**

   First, use the Pythagorean Identity $\sin^2\theta = 1 - \cos^2\theta$ to substitute for $\sin^2\theta$.

   Substituting makes: $\qquad\qquad\qquad\qquad 1 - \cos^2 - \cos\theta = \cos^2\theta$

   Collect terms: $\qquad\qquad\qquad\qquad\quad 1 - \cos^2 - \cos^2\theta - \cos\theta = 0$

   Combine like terms: $\qquad\qquad\qquad\quad\; 1 - 2\cos^2 - \cos\theta = 0$

   Rearrange terms and multiply by -1: $\quad 2\cos^2\theta + \cos\theta - 1 = 0$

   Now factor the quadratic equation: $\quad (2\cos\theta - 1)(\cos\theta + 1) = 0$

   Set each side equal to 0: $\quad \cos\theta = \dfrac{1}{2} \quad \theta = 180°, \theta = 60°, 300°$

   Therefore, $\theta = 60°, 180°, 300°$, or $\dfrac{\pi}{3}, \pi, \dfrac{5\pi}{3}$ in radians.

2. Determine whether the equation $\dfrac{y^2 + 4y}{y + 4} = y$ is an identity or a conditional equation.

   answer: **identity**

   By cross-multiplying, we find that: $\quad y^2 + 4y = y(y + 4)$

   Factor the left side to get: $\quad y(y + 4) = y(y + 4)$

   Or multiply through the right side: $\quad y^2 + 4y = y^2 + 4y$

   Ladies and gentlemen, we have ourselves an identity. Substitute whatever value you like for y, and the two sides will still be equal.

3. Solve $2\sin x - \csc x = 1$ for all possible values of x.

    answer: 90° ± (360° · n), 210° ± (360° · n), 330° ± (360° · n)

    This takes some staring at to come up with a method for solving: $2\sin x - \csc x = 1$

    The presence of sin *and* csc suggests multiplying by sin: $(2\sin x)(\sin x) - (\csc x)(\sin x) = 1 \cdot \sin x$

    Replace csc with its equivalent and cancel: $(2\sin x)(\sin x) - \left(\dfrac{1}{\sin x}\right)\left(\dfrac{\sin x}{1}\right) = \sin x$

    Multiply and simplify: $2\sin^2 x - 1 = \sin x$

    Rewrite as a quadratic equation: $2\sin^2 x - \sin x - 1 = 0$

    Factor: $(2\sin x + 1)(\sin x - 1) = 0$

    Split into two possible solutions: $2\sin x + 1 = 0$ or $\sin x - 1 = 0$

    Collect terms: $2\sin x = -1$ $\sin x = 1$

    Finish by dividing: $\sin x = -\dfrac{1}{2}$

    Find the arcsins for both values of sin: For $\sin x = 1$, $x = 90°$, but that's only the first time around. Since the problem asks for all possible values, $x = 90° \pm (360° \cdot n)$, where n is any real number.

    Now, where $\sin x = -\dfrac{1}{2}$, sin is negative in Q-III and Q-IV. Our reference angle is $30°$, which in Q-III is $210°$ and in Q-IV is $330°$, so

    $$x = 210° \pm (360° \cdot n) \text{ and } 330° \pm (360° \cdot n)$$

Work Problems
Use these problems to give yourself additional practice.

1. Without using a table or calculator find the value of $\tan(\text{arcsec}{-4})$.

2. What are the restrictions on the inverse cosecant function?

3. Solve for x's primary values: $\sin^2 x \sec x + 2\sin^2 x - \sec x - 2 = 0$.

4. Solve for all x's values: $\cos x - \sqrt{2}\sin x = 1$.

Worked Solutions

1. $-\sqrt{15}$ The angle must be in Q-II for restricted secant (and cosine) to be negative. The figure that follows shows the triangle and its reflection across the x-axis.

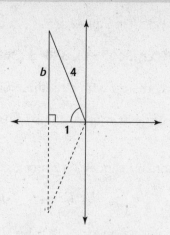

Figure for Work Problem 1.

The hypotenuse must be 4 times the adjacent side, as marked. Next, we use the Pythagorean Theorem to find the opposite side.

State the theorem: $c^2 = a^2 + b^2$

Substitute: $4^2 = 1^2 + b^2$

Square and rearrange: $b^2 = 16 - 1$

Add and take the root: $b = \sqrt{15}$

Since $b = \sqrt{15}$ $\tan = \dfrac{\sqrt{15}}{-1} = -\sqrt{15}$

2. **$x \le -1, x \ge 1,$ $y = -\dfrac{\pi}{2}$ to $\dfrac{\pi}{2}$ with $y \ne 0$**   The inverse cosecant function is derived from Sine, and has a domain restricted to x-values equal to or less than -1, or greater than or equal to 1. The range is from $y = -\dfrac{\pi}{2}$ with $y \ne 0$, since that's its asymptote. To see its graph, look back to the figure on page 253 ("The inverse cosecant function").

3. **$90°, 120°, 240°, 270°$** $\sin^2 x \sec x + 2\sin^2 x - \sec x - 2 = 0$

Factor $\sin^2 x$ out of the first two terms: $\sin^2 x (\sec x + 2) - \sec x - 2 = 0$

Factor -1 out of the last two terms: $\sin^2 x (\sec x + 2) - 1(\sec x + 2) = 0$

Now factor $\sec x + 2$ out of both terms: $(\sin^2 x - 1)(\sec x + 2) = 0$

$$\sin^2 x = 1 \quad \sec x = -2$$

Split up and solve: $\sin x = \pm 1$ $x = \sec^{-1}(-2) = 120°, 240°$

$x = \sin^{-1} \pm 1 = 90°, 270°$

4. **$0° + (360° \cdot n)$, $109.5° + (360° \cdot n)$, $250.5 + (360° \cdot n)$**

The strategy is to isolate the root on one side: $\cos x - \sqrt{2}\sin x = 1$

Rearrange the terms to do so: $\cos x - 1 = \sqrt{2}\sin x$

Next square both sides: $(\cos x - 1)^2 = (\sqrt{2}\sin x)^2$

Actually square them: $\cos^2 x - 2\cos x + 1 = 2\sin^2 x$

Use the Pythagorean Identity $\sin^2\theta = 1 - \cos^2\theta$: $\cos^2 x - 2\cos x + 1 = 2(1 - \cos^2 x)$

Clear parentheses: $\cos^2 x - 2\cos x + 1 = 2 - 2\cos^2 x$

Move everything to the left and combine: $3\cos^2 x - 2\cos x - 1 = 0$

Factor: $(3\cos x + 1)(\cos x - 1) = 0$

Set each portion equal to 0:

$3\cos x + 1 = 0$ $\qquad\qquad \cos x - 1 = 0$

$3\cos x = -1$ $\qquad\qquad \cos x = 1$

$\cos x = -\dfrac{1}{3}$ $\qquad\qquad x = \cos^{-1}(1) = 0° + (360°n)$

$x = \cos^{-1}\left(-\dfrac{1}{3}\right) = 109.5° + (360°n), 250.5 + (360°n)$

Uniform Circular Motion

Have you ever noticed that some objects seem to move with very repetitive or even rhythmic motions? The pistons in a car engine would be an example, or the drive wheels and shafts on a steam locomotive, which unfortunately is something you've probably only seen in a movie or a museum. A playground swing, a bouncing ball, and a pendulum all move with motions related to the graph of the sine curve. Understanding simple harmonic motion and circular motion can help to explain how many of these commonplace movements are related to each other.

When dealing with circular motion, we need to think in terms of radians. Consider a central angle of measure α. The length of its intercepted arc (s) may be found by multiplying the radius of the circle (r) by the size of that central angle:

$$s = r\alpha$$

Remember, α must be measured in radians. So, if a circle has a radius of 10 cm, a central angle of 2 radians intercepts an arc of:

$$s = r\alpha$$
$$s = (10)(2)$$
$$s = 20 \text{ cm}$$

The linear velocity (*v*) of a point traveling along an arc of a circle at constant speed is found by dividing the length of the arc by the time it takes to travel that arc, or:

$$v = \frac{r\alpha}{t}$$

At any given point on a circle's circumference, the linear velocity is considered as being directed along the tangent at that point, and is, therefore, perpendicular to the radius to that point.

Differing from linear velocity but easily as important in studying uniform circular motion is angular velocity. Angular velocity is represented by the Greek lower case omega (*ω*), and is the velocity of a point traveling at constant speed along the arc of a circle. The formula for angular velocity is given below:

$$\omega = \frac{\alpha}{t} \quad \text{or} \quad \frac{\text{the angle of rotation}}{\text{time}}$$

Both linear and angular velocity are considered to be positive if the rotation is counterclockwise, and negative if rotation is clockwise. Remember, the angle must be expressed in radians.

Example Problems

These problems show the answers and solutions.

1. Find the length of an arc intercepted by a central angle of 120° in a circle with radius 12 in. long.

   answer: 25.12 in.

   Remember, the angle must be in radians. $120° = \frac{2\pi}{3}$ radians

   $s = r\alpha$

   $s = (12)\left(\frac{2\pi}{3}\right)$

   $s = 8\pi$ or $8 \cdot 3.14 = 25.12$

2. The earth has a diameter at the equator of 17,820 km and rotates one complete revolution in 24 hours. What is the linear velocity of a person standing still on the equator?

   answer: 2331.45 km/hr.

   Bear in mind that the earth's radius is half the diameter, so $r = \frac{17,820}{2}$, or 8,910 km, and one full revolution is 2π.

   Now write the equation: $v = \frac{r\alpha}{t}$

   Substitute: $v = \frac{(8910)(2\pi)}{24}$

   Multiply: $v = \frac{(8910)(6.28)}{24} = \frac{55,954.8}{24}$

   And divide: $v = 2331.45$ km per hour

   Remember, that's standing still!

3. A point on a paddle wheel with an 8 foot radius makes 3 complete counterclockwise revolutions every 10 seconds. Find that point's linear and angular velocities.

   answer: 15.072 ft./sec., 1.884 ft./sec.

   First, write the equation: $v = \dfrac{r\alpha}{t}$

   Substitute: $v = \dfrac{(8)(6\pi)}{10}$

   Multiply: $v = \dfrac{(48)(3.14)}{10} = \dfrac{150.72}{10}$

   And divide: $v = 15.072$ ft. per second

   Now for the angular velocity: $\omega = \dfrac{\alpha}{t}$

   Substitute: $\omega = \dfrac{6\pi}{10}$

   Multiply: $\omega = \dfrac{6 \cdot 3.14}{10} = \dfrac{18.84}{10}$

   Divide: $\omega = 1.884$ ft. per second

   Notice the considerable difference between angular and linear velocities.

Simple Harmonic Motion

Back and forth motion such as that of a swinging pendulum or a spring that alternately compresses and expands are examples of simple harmonic motion. In some cases simple harmonic motion is the result of circular motion, and sometimes circular motion is the result of simple harmonic motion. A carousel horse's up and down motion is the result of a circular motion of a rod it's attached to. A car's circularly moving wheels are the result of the back and forth motion of the engine's pistons. In general, simple harmonic motion satisfies the equations:

$$d = a\sin Bt \text{ and/or } d = A\cos Bt$$

d is the amount of displacement or movement of the objects, *A* and *B* are constants related to the specific motion, and *t* is an interval of time. The involvement of sine and/or cosine functions validate the relationship of simple harmonic motion to circular motion.

Example Problems

These problems show the answers and solutions.

1. The end of a pendulum swings 24 inches in 6.5 seconds. If the formula for calculating a pendulum's swing is $d = K \sin 2\dfrac{\pi}{3} t$, find the value of K.

   answer: 27.84

   We've been given the displacement as 24 inches and 6.5 seconds, the time, so:

   Write the equation: $d = K\left(\sin 2\dfrac{\pi}{3} t\right)$

Substitute:

$$24 = K\sin\left(2 \cdot \frac{3.14}{3} \cdot 6.5\right)$$

Multiply:

$$24 = K\sin(13.606)$$

Isolate K:

$$K = \frac{24}{\sin 13.606}$$

Find sin13.325 (remember radians mode):

$$K = \frac{24}{0.862}$$

Divide and round:

$$K = 27.84$$

2. The displacement of a spring is given by the formula $d = A\cos\alpha t$, where the initial displacement A, and d are expressed in centimeters, t in seconds, and $\alpha = 6$. Find d when $t = \pi$ and $A = 12$ cm.

answer: $d \approx 12$ cm

Write the equation: $d = A\cos\alpha t$

Substitute: $d = 12\cos(6\pi)$

Multiply: $d = 12\cos(6 \cdot 3.14) = 12\cos(18.84)$

Find the cosine: $d \approx 12\,(1)$

And multiply again: $d \approx 12$ cm

3. The driving end of a piston on a steam engine traverses d inches in 6 seconds. If the formula for calculating a piston's period is $d = K\sin\frac{\pi}{4}t$, and K is 8, find the value of d.

answer: 8 inches

We've been given the constant as 8 and 6 seconds, the time, so:

Write the equation: $d = K\sin\frac{\pi}{4}t$

Substitute: $d = 8\sin\left(\frac{3.14}{\underset{2}{4}} \cdot \overset{3}{6}\right)$

Multiply: $d = 8\sin(4.71)$

Find sin9.42 (remember radians mode): $d \approx 8(-1)$

Multiply: $d \approx 8$ in.

The negative indicates that the displacement was in a clockwise direction, but we drop it since distances aren't negative.

Work Problems

Use these problems to give yourself additional practice.

1. The moon always keeps the same face toward the earth. It is roughly 250,000 miles from the earth and makes one complete orbit in about 30 days. What is the linear velocity of a mountain on the moon in miles per hour?

2. An automobile's crankshaft traverses 88 feet in 1 second. If the formula for calculating a crankshaft's displacement is $d = K \sin \frac{\pi}{3} t$, find the value of K.

3. A seat on a Ferris wheel with a 30 foot radius makes 2 complete counterclockwise revolutions every 90 seconds. Find that seat's linear and angular velocities.

4. The displacement of a spring is given by the formula $d = A \cos \alpha t$, where the initial displacement A, and d are expressed in inches, t in seconds, and $\alpha = 12$. Find d when $t = 3\pi$ and $A = 10$ in.

Worked Solutions

1. **2,181 mph**   Bear in mind that one full revolution is 2π. 250,000 miles is roughly the radius of the orbit.

   Now write the equation: $v = \dfrac{r\alpha}{t}$

   Substitute: $v = \dfrac{(250,000)(2\pi)}{30}$

   Multiply: $v = \dfrac{(250,000)(6.28)}{30} = \dfrac{1,570,000}{30}$

   And divide: $v = 52,333.333$ miles per day

   But we want miles per hour: $v = \dfrac{52,333.333}{24} = 2180.56$

   Let's call it 2,181 mph. Talk about moving mountains!

2. **4,888.9**

   We've been given the displacement as 88 feet and 1 second as the time, so:

   Write the equation: $d = K \sin \frac{\pi}{3} t$

   Substitute: $88 = K \sin\left(\frac{3.14}{3} \cdot 1\right)$

   Multiply: $88 = K \sin(1.05)$

   Isolate K: $K = \dfrac{88}{\sin(1.05)}$

   Find sin1.05 (remember radians mode): $K = \dfrac{88}{.018}$

   Divide: $K = 4888.9$ in.

3. **4.2 ft./sec., 0.14 ft./sec.**

First, write the equation: $v = \dfrac{r\alpha}{t}$

Substitute: $v = \dfrac{(30)(4\pi)}{90}$

Multiply: $v = \dfrac{(30)(4)(3.14)}{90} = \dfrac{376.8}{90}$

And divide: $v = \dfrac{376.8}{90} = 4.2$ per second

Now for the angular velocity: $\omega = \dfrac{\alpha}{t}$

Substitute and simplify: $\omega = \dfrac{4\pi}{90} = \dfrac{2\pi}{45}$

Multiply: $\omega = \dfrac{2 \cdot 3 \cdot 14}{45} = \dfrac{6.28}{45}$

Divide: $\omega = 0.14$ ft. per second

4. **$d \approx 9.98$ in**

Write the equation: $d = A\cos\alpha t$

Substitute: $d = 10\cos(12 \cdot 3\pi)$

Multiply: $d = 10\cos(12 \cdot 3 \cdot 3.14) = 10\cos(113.04)$

Find the cosine: $d \approx 10(0.998)$

And multiply again: $d \approx 9.98$ in.

Chapter Problems and Solutions

Problems

Solve these problems for more practice applying the skills from this chapter. Worked out solutions follow problems.

1. What restriction is placed upon the domain of the restricted sine function?

2. Find the exact value of $\mathrm{Cos}\left(-\dfrac{1}{2}\right)$.

3. Find the exact value of $\mathrm{Sin}\left(-\dfrac{\sqrt{2}}{2}\right)$.

4. Find the exact value of $\cos(\mathrm{Cos}^{-1}0.27)$.

5. Evaluate $\tan(\arcsin\frac{7}{25})$ without using a calculator or table of functions.

6. Find the value of $\text{Tan}^{-1}(1.6)$ to the nearest tenth of a degree.

7. Find the exact value of $\text{Tan}^{-1}(-1)$.

8. Without using a calculator or table of trigonometric functions, determine the precise value of $\sin\left[\text{Sec}^{-1}\left(\frac{40}{24}\right)\right]$.

9. Without using a calculator or table of trigonometric functions, determine the precise value of $\cos\left[\text{Sec}^{-1}\left(\frac{85}{36}\right)\right]$.

10. Without using a table or calculator find the value of $\tan(\text{arcsec}-7)$.

11. Find the exact values of θ in the equation $\cos^2\theta - \sin\theta = \sin^2\theta$ for $0 \leq \theta \leq 2\pi$.

In 12–15, prove that each is an identity.

12. $(\cos\phi + \sin\phi)^2 = 2\cos\phi\,\sin\phi + 1$

13. $\csc^2 y - \sin^2 y - \cos^2 y = \cot^2 y$

14. $(\sec x + 1)(\sec x - 1) = \tan^2 x$

15. $1 - \sin 2x = (\sin x - \cos x)^2$

16. The end of a pendulum swings 6 inches. If the formula for calculating a pendulum's swing is $d = K\sin\frac{\pi}{6}t$, and the value of K is 12, find t to the nearest second.

17. The displacement of a spring is given by the formula $d = A\cos\alpha t$, where the initial displacement A, and d are expressed in centimeters, t in seconds, and $\alpha = \frac{3\pi}{2}$. Find d when $t = 8$ and $A = 16$ cm.

18. Find the length of an arc intercepted by a central angle of $300°$ in a circle with radius 24 mm long.

19. A point on a water wheel with a 12-foot diameter makes 5 complete counterclockwise revolutions every 2 minutes. Find that point's linear and angular velocities.

Answers and Solutions

1. **Answer:** $-\dfrac{P}{2} \leq x \leq \dfrac{P}{2}$ (See the following figure.)

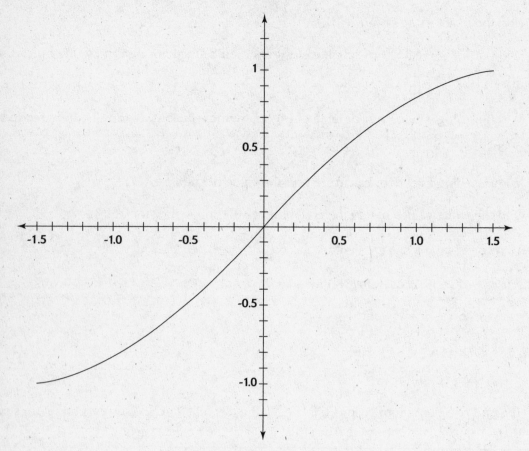

Restricted Sine: $y = \text{Sin } x$.

2. **Answer: 120° or $\dfrac{2\pi}{3}$** Following the rationale one uses to solve trigonometric inverses is not the same as one would use on an algebraic equation.

   Suppose $y = \text{Cos}\left(-\dfrac{1}{2}\right)$. That means that $\cos y = y = -\dfrac{1}{2}$ where $0 \leq y \leq \pi$.

   We should recognize $\dfrac{1}{2}$ to be the cosine of 60°, or $\dfrac{\pi}{3}$ radians. Since the cosine is negative, it must be in the second quadrant. Subtract $180° - 60° = 120°$. In radians, $\pi - \dfrac{\pi}{3} = \dfrac{2\pi}{3}$.

3. **Answer: 315°, or $\dfrac{7\pi}{4}$** Suppose $y = \text{Sin}\left(\dfrac{1}{2}\right)$. That means that $\sin y = -\dfrac{\sqrt{2}}{2}$ where $-1 \leq y \leq 1$. $\dfrac{\sqrt{2}}{2}$ is the sin of 45°, or $\dfrac{\pi}{4}$. But the given figure was negative. Restricted sine is negative in Q-IV, so the solution is $360° - 45° = 315°$, or $\dfrac{7\pi}{4}$.

4. **Answer: 0.27** Use the cosine-inverse cosine identity: $\cos(\text{Cos}^{-1} x) = x \quad -1 \leq x \leq 1$

   Therefore: $\cos(\text{Cos}^{-1} 0.27) = 0.27$

5. ***Answer:*** $\frac{7}{24}$ If you draw a right triangle in standard position, you'll find that it is a triangle whose sine is $\frac{7}{25}$. That means its opposite side is 7 and its hypotenuse is 25.

Apply the Pythagorean Theorem to get the adjacent side: $a^2 = c^2 - b^2$

Substitute: $a^2 = 25^2 - 7^2$

Square: $a^2 = 625 - 49$

Subtract: $a^2 = 576$

Take the square root: $a = 24$

So, if the adjacent side is 24, $\tan = \dfrac{\text{opposite}}{\text{adjacent}} = \dfrac{7}{24}$. There's a new Pythagorean Triple for you, 7-24-25.

6. ***Answer:*** **58°**   The question asks for the angle whose tangent is 1.6. Using a calculator or the table (the latter of which requires some interpolation), we find it is 57.9964°. To the nearest tenth, that's 58°.

7. ***Answer:*** **315°**   A tangent of 1 suggests an isosceles right triangle, hence 45° angles. But wait. The tangent was −1. The range of Tan^{-1}, you'll recall, is $-\frac{\pi}{2} < y < \frac{\pi}{2}$. That means the first and fourth quadrants. Tangent is negative in Q-IV, so 45° is the reference angle, but we will not leave it as −45°, but will add one rotation and make it −45° + 360° = 315°.

8. ***Answer:*** $\frac{4}{5}$ Suppose we set the bracketed expression equal to the angle, ϕ.

Then: $\phi = \text{Sec}\left(\dfrac{40}{24}\right)$

That means: $\text{Sec}\,\phi = \dfrac{40}{24}$, where $0 \leq \phi \leq \pi,\ \phi \neq \dfrac{\pi}{2}$

If you draw a diagram showing a triangle in standard position in Q-I, you should have the adjacent side and hypotenuse. We can use the Pythagorean Theorem to find the opposite side:

State the theorem: $a^2 + b^2 = c^2$

Substitute: $a^2 + 24^2 = 40^2$

Square: $a^2 + 576 = 1600$

Collect terms: $a^2 = 1600 - 576$

Subtract and take the root: $a = \sqrt{1024} = 32$

Therefore: $\sin\left[\text{Sec}^{-1}\left(\dfrac{40}{24}\right)\right] = \sin\phi$

And: $\sin\left[\text{Sec}^{-1}\left(\dfrac{40}{24}\right)\right] = \dfrac{a}{40}$

And finally: $= \dfrac{32}{40} = \dfrac{4}{5}$

9. ***Answer:*** $\dfrac{36}{85}$ The cosine is the inverse of the secant, so the angle whose secant is $\dfrac{85}{36}$ must have a cosine of $\dfrac{36}{85}$.

10. ***Answer:*** $4\sqrt{3}$ **if in Q-III or** $-4\sqrt{3}$ **if in Q-II**   The angle must be in Q-II or Q-III for secant (and cosine) to be negative. The hypotenuse must be 7 times the adjacent side, for the arcsecant to have an absolute value of 7. Next, use the Pythagorean Theorem to find the opposite side.

State the theorem: $c^2 = a^2 + b^2$

Substitute: $7^2 = 1^2 + b^2$

Square and rearrange: $b^2 = 49 - 1$

Add and take the root: $b = \sqrt{48} = \sqrt{16 \cdot 3} = 4\sqrt{3}$

Since $b = 4\sqrt{3}$, $\tan = \dfrac{4\sqrt{3}}{1} = 4\sqrt{3}$ if in Q-III or $-4\sqrt{3}$ if in Q-II.

11. ***Answer:*** $\theta = 30°, 150°, 270°$ *or* $\dfrac{\pi}{6}, \dfrac{5\pi}{6}, \dfrac{11\pi}{6}$ **in radians.**

First, use the Pythagorean Identity $\cos^2\theta = 1 - \sin^2\theta$ to substitute for $\cos^2\theta$.

Substituting makes: $1 - \sin^2\theta - \sin\theta = \sin^2\theta$

Collect terms: $1 - \sin^2\theta - \sin^2\theta - \sin\theta = 0$

Combine like terms: $1 - 2\sin^2 - \sin\theta = 0$

Rearrange terms and multiply by -1: $2\sin^2 + \sin\theta - 1 = 0$

Now factor the quadratic equation: $(2\sin\theta - 1)(\sin\theta + 1) = 0$

Set each side equal to 0: $\sin\theta \dfrac{1}{2}$ $\theta = 270°$

Therefore, $\theta = 30°, 150°, 270°$ or $\dfrac{\pi}{6}, \dfrac{5\pi}{6}, \dfrac{11\pi}{12}$ in radians.

12. ***Answer:*** $1 = 1$

First, we'll rewrite the equation: $(\cos\phi + \sin\phi)^2 = 2\cos\phi \sin\phi + 1$

Clear the parentheses, using FOIL: $\cos^2\phi + 2\cos\phi \sin\phi + \sin^2\phi = 2\cos\phi \sin\phi + 1$

These two terms cancel: $\cos^2\phi + \underline{2\cos\phi\sin\phi} + \sin^2\phi = \underline{2\cos\phi\sin\phi} + 1$

That leaves a possibly recognizable: $\cos^2\phi + \sin^2\phi = 1$

That's the Pythagorean Identity: $\sin^2\theta + \cos^2\theta = 1$, therefore: $1 = 1$

13. ***Answer:*** $1 = 1$

Write the equation: $\csc^2 y - \sin^2 y - \cos^2 y = \cot^2 y$

Well for openers, $1 + \cot^2\theta = \csc^2\theta$, so: $1 + \cot^2 y - \sin^2 y - \cos^2 y = \cot^2 y$

Next, subtract $\cot^2 y$ from both sides: $1 - \sin^2 y - \cos^2 y = 0$

Next add $\sin^2 y + \cos^2 y$ to both sides: $1 = \sin^2 y + \cos^2 y$

And that's a Pythagorean identity: $\sin^2\theta + \cos^2\theta = 1$, so $1 = 1$

14.   ***Answer:** 1 = 1*

     Write the equation: $(\sec x + 1)(\sec x - 1) = \tan^2 x$

     Clear parentheses: $\sec^2 x - 1 = \tan^2 x$

     That's the identity: $\tan^2\theta = \sec^2\theta - 1$, so $1 = 1$

15.   ***Answer:** 1 = 1*

     First, we'll rewrite the equation: $1 - \sin 2x = (\sin x - \cos x)^2$

     Clear the parentheses, using FOIL: $1 - \sin 2x = \sin^2 x - 2\sin x \cos x + \cos^2 x$

     Rearrange terms: $1 - \sin 2x = \sin^2 x = \sin^2 x + \cos^2 x - 2\sin x \cos x$

     Recognizing that $\cos^2\phi + \sin^2\phi = 1$,
     replace them: $1 - \sin 2x = 1 - 2\sin x \cos x$

     Finally, since $\sin 2\theta = 2\sin\theta\cos\theta$: $1 - \sin 2x = 1 - \sin 2x$, so $1 = 1$

16.   ***Answer:** 1 sec.   t = 1*

     We've been given the displacement as 6 inches and $K = 12$. To find the time:

     Write the equation: $d = K \sin\dfrac{\pi}{6} t$

     Substitute: $6 = 12\sin\left(\dfrac{\pi}{6}\right)(t)$

     Divide by $\dfrac{1}{2}$: $\dfrac{1}{2} = \sin\dfrac{\pi}{6} t$

     Take Sin^{-1}: $\mathrm{Sin}^{-1}\left(\dfrac{1}{2}\right) = \mathrm{Sin}^{-1}\left(\sin\dfrac{\pi}{6} t\right)$

     Simplify: $\dfrac{\pi}{6} = \dfrac{\pi}{6} t$

     Divide by $\dfrac{\pi}{6}$: $1 = t$

17.   ***Answer:** 16 in.   d ≈ 16 in.*

     Write the equation: $d = A \cos \alpha t$

     Substitute: $d = 16\cos\left(\dfrac{3\pi}{2}\right)(8) = 16\cos(3\pi)(4)$

     Multiply: $d = 16\cos(12 \cdot 3.14) = 12\cos(37.68)$

Find the cosine: $d \approx 16(1)$

And multiply again: $d \approx 16$ in.

18. **Answer: 125.6 mm**

Remember, the angle must be in radians: $300° = \frac{5\pi}{3}$ radians

Write the equation: $s = r\alpha$

Substitute and simplify: $s = (24)\left(\frac{5\pi}{3}\right) = (24)\left(\frac{5\pi}{31}\right) = (8)(5\pi)$

Multiply: $s = 40\pi$ or $40 \cdot 3.14 = 125.6$ mm

19. **Answer: 1.57 ft./sec., 0.26 ft./sec.**

First, write the equation: $v = \frac{r\alpha}{t}$

Substitute using radius and seconds: $v = \frac{(6)(10\pi)}{120}$

Multiply: $v = \frac{(60)(3.14)}{120} = \frac{188.4}{120}$

And divide: $v = 1.57$ ft. per second

Now for the angular velocity: $\omega = \frac{\alpha}{t}$

Substitute: $\omega = \frac{10\pi}{120}$

Multiply: $\omega = \frac{10 \cdot 3.14}{120} = \frac{31.4}{120}$

Divide: $\omega = 0.26$ ft. per second

Supplemental Chapter Problems

Solve these problems for even more practice applying the skills from this chapter. The Answer section will direct you to where you need to review.

Problems

1. What restriction is placed upon the domain of the restricted cosine function?

2. Find the exact value of $\mathrm{Cos}^{-1}\left(-\frac{\sqrt{3}}{2}\right)$.

3. Find the exact value of $\mathrm{Sin}^{-1}\left(-\frac{1}{2}\right)$.

4. Find the exact value of $\cos(\mathrm{Cos}^{-1}0.62)$.

5. Evaluate $\tan(\arcsin\frac{7}{25})$ without using a calculator or table of functions.

6. Find the value of $\text{Tan}^{-1}(3)$ to the nearest tenth of a degree.

7. Find the exact value of $\tan\left(\text{Tan}\dfrac{3}{3\sqrt{3}}\right)$.

8. Without using a calculator or table of trigonometric functions, determine the precise value of $\sin\left[\text{Sec}^{-1}\left(\dfrac{17}{8}\right)\right]$.

9. Without using a calculator or table of trigonometric functions, determine the precise value of $\tan\left[\text{Sec}^{-1}\left(\dfrac{34}{16}\right)\right]$.

10. Without using a table or calculator find the value of $\tan(\text{arcsec}-9)$.

For problems 11–13, prove that each trigonometric equation is an identity.

11. $\dfrac{\tan a + \tan b}{\tan a - \tan b} = \dfrac{\sin(a+b)}{\sin(a-b)}$.

12. $\dfrac{\sin z}{1 + \cos z} = \tan\dfrac{1}{2}z$.

13. $\dfrac{\sec y - 1}{2\sec y} = \sin^2\dfrac{y}{2}$.

For problems 14–15, find all positive values of the angle less than or equal to 360° to the nearest degree.

14. $5\tan x + 14 = \tan^2 x$.

15. $7\sin\theta + 3\cos^2\theta = 5$.

16. Find the length of an arc intercepted by a central angle of 210° in a circle with diameter 8 feet long.

17. A point on the equator of a rolling ball with a 16 cm radius makes 3 complete counterclockwise revolutions every 8 seconds. Find that point's linear and angular velocities.

18. The end of a pendulum swings 30 inches in 5.4 seconds. If the formula for calculating a pendulum's swing is $d = K\sin\dfrac{\pi}{3}t$, find the value of K.

19. The driving end of a piston on a steam engine traverses d inches in 6.9 seconds. If the formula for calculating a piston's period is $d = K\sin\dfrac{\pi}{3}t$, and K is 48, find the value of d.

Answers

1. $0 \leq x \leq \pi$ (Restricting Functions, p. 242)

2. $150°$ or $\dfrac{5\pi}{6}$ (Inverse Sine and Cosine, p. 244)

3. $330°$ or $\dfrac{11\pi}{6}$ (Inverse Sine and Cosine, p. 244)

4.   0.62   (Inverse Sine and Cosine, p. 244)

5. $\frac{7}{24}$ (Inverse Tangent, p. 249)

6.   71.6°   (Inverse Tangent, p. 249)

7.   30° or $\frac{\pi}{6}$ (Inverse Tangent, p. 249)

8. $\frac{15}{17}$ (Inverses of Reciprocal Functions, p. 253)

9. $\frac{30}{16}$ (Inverses of Reciprocal Functions, p. 253)

10. $-4\sqrt{5}$ (Inverses of Reciprocal Functions, p. 253)

11.   1 = 1   (Trigonometric Equations, p. 257)

12.   1 = 1   (Trigonometric Equations, p. 257)

13.   1 = 1   (Trigonometric Equations, p. 257)

14.   82°, 117°, 262° and 297°   (Trigonometric Equations, p. 257)

15.   19°, 161°   (Trigonometric Equations, p. 257)

16.   14.67 ft.   (Uniform Circular Motion, p. 261)

17.   37.7 cm/sec, 2.4 cm/sec   (Uniform Circular Motion, p. 261)

18.   −50.68   (Simple Harmonic Motion, p. 263)

19.   234.2 in.   (Simple Harmonic Motion, p. 263)

Customized Full-Length Exam

Problems

1. In which quadrant is the terminal side of a 105° angle in standard position?

Answer: Second

If you answered correctly, go to problem 3.
If you answered incorrectly, go to problem 2.

2. In which angle is the terminal side of a −315° angle?

Answer: Fourth

If you answered correctly, go to problem 3.
If you answered incorrectly, review "Angles and Quadrants" on page 21.

3. Is an angle measuring 210° coterminal with an angle measuring 950°?

Answer: No

If you answered correctly, go to problem 5.
If you answered incorrectly, go to problem 4.

4. What is the lowest possible positive degree measure for an angle that is coterminal with one of −1760°?

Answer: 40°

If you answered correctly, go to problem 5.
If you answered incorrectly, review "Coterminal Angles" on page 25.

5. A right triangle has side *a* opposite acute angle *A* of length 5 cm and side *b* opposite acute angle *B* of 12 cm. What is Angle *A*'s cosecant?

Answer: $\frac{13}{5}$

If you answered correctly, go to problem 7.
If you answered incorrectly, go to problem 6.

6. A right triangle has side *a* opposite acute angle *A* of length 5 cm and side *b* opposite acute angle *B* of 12 cm. What is Angle *B*'s cotangent?

Answer: $\dfrac{5}{12}$

If you answered correctly, go to problem 7.
If you answered incorrectly, review "Trigonometric Functions of Acute Angles" on page 28.

7. What is the reciprocal of the function secant?

Answer: cosine

If you answered correctly, go to problem 9.
If you answered incorrectly, go to problem 8.

8. Which function may be written as $\dfrac{1}{\sin\theta}$?

Answer: $\csc\theta$

If you answered correctly, go to problem 9.
If you answered incorrectly, review "Reciprocal Trigonometric Functions" on page 29.

9. Express $\tan\phi$ in terms of $\sin\phi$ and $\cos\phi$.

Answer: $\dfrac{\sin\phi}{\cos\phi}$

If you answered correctly, go to problem 11.
If you answered incorrectly, go to problem 10.

10. Express $\sin^2\theta$ in terms of $\cos\theta$.

Answer: $1 - \cos^2\theta$

If you answered correctly, go to problem 11.
If you answered incorrectly, review "Introducing Trigonometric Identities" on page 32.

11. Which of the following is the trigonometric cofunction of $\sin A = \dfrac{a}{c}$?

   a. $\sin B = \dfrac{b}{c}$

   b. $\cos A = \dfrac{b}{c}$

   c. $\cos B = \dfrac{a}{c}$

Answer: c

If you answered correctly, go to problem 13.
If you answered incorrectly, go to problem 12.

12. Which of the following is the trigonometric cofunction of $\sec A = \dfrac{c}{b}$?

    a. $\sec B = \dfrac{c}{a}$

    b. $\csc B = \dfrac{c}{b}$

    c. $\cos B = \dfrac{b}{c}$

Answer: b

If you answered correctly, go to problem 13.
If you answered incorrectly, review "Trigonometric Cofunctions" on page 34.

13. One leg of an isosceles right triangle is 8 cm long. How long is the hypotenuse?

Answer: $8\sqrt{2}$ cm

If you answered correctly, go to problem 15.
If you answered incorrectly, go to problem 14.

14. The shortest leg of a 30-60-90 triangle is 9 inches long. How long is the hypotenuse?

Answer: 18 in.

If you answered correctly, go to problem 15.
If you answered incorrectly, review "Two Special Triangles" on page 35.

15. What is the sign of tan 250°?

Answer: positive

If you answered correctly, go to problem 17.
If you answered incorrectly, go to problem 16.

16. What is the sign of sec 300°?

Answer: positive

If you answered correctly, go to problem 17.
If you answered incorrectly, review "Functions of General Angles" on page 38.

17. Without using a table or a calculator, find the cosine of 150°?

Answer: $-\dfrac{\sqrt{3}}{2}$

If you answered correctly, go to problem 19.
If you answered incorrectly, go to problem 18.

18. Without using a table or a calculator, find the sine of 330°?

Answer: $-\dfrac{1}{2}$

If you answered correctly, go to problem 19.
If you answered incorrectly, review "Reference Angles" on page 41.

19. Using a table of trigonometric functions, find tan46.8° to the nearest thousandth.

Answer: 1.065

If you answered correctly, go to problem 21.
If you answered incorrectly, go to problem 20.

20. Using a table of trigonometric functions, find cos38.5° to the nearest thousandth.

Answer: 0.783

If you answered correctly, go to problem 21.
If you answered incorrectly, review "Interpolation" on page 45.

21. What instrument is used to find degree measure of angles?

Answer: a protractor

If you answered correctly, go to problem 23.
If you answered incorrectly, go to problem 22.

22. How useful is a protractor for finding minute divisions of an angle?

Answer: Useless

If you answered correctly, go to problem 23.
If you answered incorrectly, review "Understanding Degree Measure" on page 59.

23. What is the measure of an arc of 1 radian on the circumference of a circle?

Answer: The length of a radius or $\dfrac{C}{2\pi}$

If you answered correctly, go to problem 25.
If you answered incorrectly, go to problem 24.

24. How many radians long is the circumference of a circle?

Answer: 2π

If you answered correctly, go to problem 25.
If you answered incorrectly, review "Understanding Radians " on page 59.

25. How many degrees is $\dfrac{\pi}{6}$ radians?

Answer: 30°

If you answered correctly, go to problem 27.
If you answered incorrectly, go to problem 26.

26.   Express 120° in radians.

Answer: $\dfrac{2\pi}{3}$

If you answered correctly, go to problem 27.
If you answered incorrectly, review "Relationships Between Degrees and Radians" on page 60.

27.   In a unit circle, what is the value of the sine of the central angle?

Answer: y

If you answered correctly, go to problem 29.
If you answered incorrectly, go to problem 28.

28.   In a unit circle, what is the value of the tangent of a central angle?

Answer: $\dfrac{y}{x}$

If you answered correctly, go to problem 29.
If you answered incorrectly, review "The Unit Circle and Circular Functions" on page 62.

29.   What values of x in the domain of the sine function between 0 and π have a value equal to an absolute value of 1?

Answer: $\dfrac{\pi}{2}$

If you answered correctly, go to problem 31.
If you answered incorrectly, go to problem 30.

30.   What values of x in the domain of the cosine function between 0 and π have a range equal to 0?

Answer: $\dfrac{\pi}{2}$

If you answered correctly, go to problem 31.
If you answered incorrectly, review "Domain vs. Range" on page 63.

31.   What is the simplest expression for $\sin(\theta + 2n\pi)$?

Answer: $\sin\theta$

If you answered correctly, go to problem 33.
If you answered incorrectly, go to problem 32.

32.   What is the simplest expression for $\cos(\theta + 2n\pi)$

Answer: $\cos\theta$

If you answered correctly, go to problem 33.
If you answered incorrectly, review "Periodic Functions" on page 66.

33. When the graph of sine is at (0,0), where is the graph of cosine?

Answer: (0,1)

If you answered correctly, go to problem 35.
If you answered incorrectly, go to problem 34.

34. How much out of phase are the sine and cosine graphs?

Answer: 90° or $\frac{\pi}{2}$ radians

If you answered correctly, go to problem 35.
If you answered incorrectly, review "Graphing Sine and Cosine" on page 67.

35. What is the amplitude of the cosine function?

Answer: 1

If you answered correctly, go to problem 37.
If you answered incorrectly, go to problem 36.

36. What is the amplitude of the tangent function?

Answer: ∞

If you answered correctly, go to problem 37.
If you answered incorrectly, review "Vertical Displacement and Amplitude" on page 70.

37. Find the frequency and phase shift of $y = \sin(4x - 5)$

Answer: 4, −5

If you answered correctly, go to problem 39.
If you answered incorrectly, go to problem 38.

38. Find the frequency and phase shift of $y = 4 - 3\cos(3x - 8)$.

Answer: 3, −8

If you answered correctly, go to problem 39.
If you answered incorrectly, review "Frequency and Phase Shift " on page 74.

39. When is tangent undefined?

Answer: when cosine = 0

If you answered correctly, go to problem 41.
If you answered incorrectly, go to problem 40.

40. What value does tangent approach when it's undefined?

Answer: ∞

If you answered correctly, go to problem 41.
If you answered incorrectly, review "Graphing Tangent" on page 77.

41. What do we call a line that a graph can approach but never actually reach?

Answer: an asymptote

If you answered correctly, go to problem 43.
If you answered incorrectly, go to problem 42.

42. When the tangent curve comes back from being undefined where does it seem to come from?

Answer: $-\infty$

If you answered correctly, go to problem 43.
If you answered incorrectly, review "Asymptotes" on page 77.

43. How is the cotangent's graph different from the tangent's?

Answer: Tangent runs from $-\infty$ to ∞; cotangent runs from ∞ to $-\infty$, moving from left to right.

If you answered correctly, go to problem 45.
If you answered incorrectly, go to problem 44.

44. Name two differences between the sine and cosecant functions' graphs.

Answer: Cosecant is discontinuous while sine is not; cosecant goes everywhere where sine does not.

If you answered correctly, go to problem 45.
If you answered incorrectly, review "Graphing the Reciprocal Functions" on page 78.

45. In right triangle PQR, $m\angle P = 60°$ and hypotenuse $q = 12$. Solve the triangle.

Answer: $m\angle Q = 90°$, $m\angle R = 30°$, $p = 6\sqrt{3}$, $r = 6$

If you answered correctly, go to problem 47.
If you answered incorrectly, go to problem 46.

46. In right triangle STU, $m\angle S = 25°$ and U is the right angle. $t = 10$. Solve the triangle.

Answer: $m\angle T = 65°$, $s = 5$, $u = 11$.

If you answered correctly, go to problem 47.
If you answered incorrectly, review "Finding Missing Parts of Right Triangles" on page 93.

47. The foot of a ladder is 8 feet away from the bottom of a wall. The ladder forms a 60° angle of elevation with the ground. How long is the ladder?

Answer: 16 ft.

If you answered correctly, go to problem 49.
If you answered incorrectly, go to problem 48.

48. The pilot of an airplane flying at 52,000 feet spots another plane flying at 48,000 feet. The angle of depression to the second plane is 45°. What is the length of the pilot's line of sight to the second plane?

Answer: 5658 ft.

If you answered correctly, go to problem 49.
If you answered incorrectly, review "Angles of Elevation and Depression" on page 95.

49. Solve triangle ABC if $m\angle A = 60°$, $m\angle C = 50°$ and $b = 16$. Express the solution to the nearest integer.

Answer: $m\angle B = 70°$, $a = 15$, $c = 13$

If you answered correctly, go to problem 51.
If you answered incorrectly, go to problem 50.

50. Given triangle GHI find $m\angle H$ if $m\angle G = 35°$, $g = 10$ in. and $h = 16$ in. Express the solution to the nearest degree.

Answer: 67°

If you answered correctly, go to problem 51.
If you answered incorrectly, review "The Law of Sines" on page 100.

51. In $\triangle ABC$, $a = 10$ $b = 14$, $c = 12$. Find $\angle B$. Express the solution to the nearest degree.

Answer: 79°

If you answered correctly, go to problem 53.
If you answered incorrectly, go to problem 52.

52. In $\triangle CDE$, $c = 8$, $d = 10$, $e = 13$. Find $\angle$s C, D, and E to the nearest degree.

Answer: $\angle C = 38°$, $\angle D = 50°$, $\angle E = 92°$

If you answered correctly, go to problem 53.
If you answered incorrectly, review "The Law of Cosines" on page 105.

53. In $\triangle MNO$, $m = 86$, $n = 62$, and $o = 41$. Find the measure of the angles to the nearest whole degree.

Answer: $m\angle M \approx 112°$, $m\angle N \approx 42°$, $m\angle O \approx 26°$

If you answered correctly, go to problem 55.
If you answered incorrectly, go to problem 54.

54. In $\triangle ABC$, $c = 12$, $b = 12$, and $A = 70°$. Solve the triangle rounding all lengths and angles to the nearest whole units.

Answer: $m\angle C = 55°$, $m\angle B = 55°$, $c \approx 14$

If you answered correctly, go to problem 55.
If you answered incorrectly, review "Solving General Triangles (SSS and SAS)" on page 108.

55. Find the number of triangles that can be formed if $\angle C = 90°$, $c = 45$, and $b = 38$.

Answer: One

If you answered correctly, go to problem 57.
If you answered incorrectly, go to problem 56.

56. If $s = 12$, $r = 8$, $\angle S = 50°$, can $\triangle RST$ be one, two, or no triangle(s)?

Answer: One

If you answered correctly, go to problem 57.
If you answered incorrectly, review "SSA, The Ambiguous Case" on page 111.

57. Specific dimensions of triangle *MNO* are $\angle N = 45°$, $m = 8$ m, $o = 6$ m. Find its area to the nearest square unit.

Answer: 17 m^2

If you answered correctly, go to problem 59.
If you answered incorrectly, go to problem 58.

58. Some dimensions of triangle *EFG* are $e = 16$ cm, $f = 22$ cm, $\angle G = 30°$. Find its area to the nearest square unit.

Answer: 88 cm^2

If you answered correctly, go to problem 59.
If you answered incorrectly, review "Area for SAS" on page 116.

59. Certain dimensions of triangle *ABC* are $b = 12$ cm, $\angle A = 50°$, $\angle C = 70°$. Find its area to the nearest square cm.

Answer: 60 cm^2

If you answered correctly, go to problem 61.
If you answered incorrectly, go to problem 60.

60. Certain dimensions of triangle *RST* are $\angle R = 55°$, $\angle S = 65°$, $t = 10$ in. Find its area to the nearest square inch.

Answer: 43 in.^{2}

If you answered correctly, go to problem 61.
If you answered incorrectly, review "Area for ASA or SAA " on page 117.

61. Use Heron's Formula to find the areas of the triangle with sides $c = 10$ cm, $d = 24$ cm, $e = 26$ cm.

Answer: 120 cm^2

If you answered correctly, go to problem 63.
If you answered incorrectly, go to problem 62.

62. Use Heron's Formula to find the areas of the triangle with sides $a = 9$ in., $b = 12$ in., $c = 15$ in.

Answer: 54 in.^{2}

If you answered correctly, go to problem 63.
If you answered incorrectly, review "Heron's Formula (SSS)" on page 119.

63. Write the reciprocal identity for $\sec\theta$.

Answer: $\sec\theta = \dfrac{1}{\cos\theta}$

If you answered correctly, go to problem 65.
If you answered incorrectly, go to problem 64.

64. Write the reciprocal identity for $\dfrac{1}{\sec\theta}$.

Answer: $\dfrac{1}{\sec\theta} = \cos\theta$

If you answered correctly, go to problem 65.
If you answered incorrectly, review "Reciprocal Identities" on page 137.

65. Write the ratio identity for tangent θ.

Answer: $\tan\theta = \dfrac{\sin\theta}{\cos\theta}$

If you answered correctly, go to problem 67.
If you answered incorrectly, go to problem 66.

66. Write the ratio identity for cotangent θ.

Answer: $\cot\theta = \dfrac{\cos\theta}{\sin\theta}$

If you answered correctly, go to problem 67.
If you answered incorrectly, review "Ratio Identities" on page 138.

67. State the cofunction identity for $\sin A$ as a ratio in a triangle with a right angle at C.

Answer: $\sin A = \dfrac{a}{c} = \cos B$

If you answered correctly, go to problem 69.
If you answered incorrectly, go to problem 68.

68. State the cofunction identity for secant θ as a complement.

Answer: $\sec\theta = \csc(90° - \theta)$

If you answered correctly, go to problem 69.
If you answered incorrectly, review "Cofunction Identities" on page 139.

69. Write the identity for $\sin(-\theta)$.

Answer: $\sin(-\theta) = -\sin\theta$

If you answered correctly, go to problem 71.
If you answered incorrectly, go to problem 70.

70. Write the identity for $-\csc\theta$.

Answer: $\csc(-\theta) = -\csc\theta$

If you answered correctly, go to problem 71.
If you answered incorrectly, review "Identities for Negatives" on page 139.

71. State one of the Pythagorean identities that relates sine and cosine.

Answer: $\sin^2\theta + \cos^2\theta = 1$ or $\sin^2\theta = 1 - \cos^2\theta$ or $\cos^2\theta = 1 - \sin^2\theta$

If you answered correctly, go to problem 73.
If you answered incorrectly, go to problem 72.

72. State one of the Pythagorean identities relating tangent and secant.

Answer: $\sec^2\theta - \tan^2\theta = 1$ or $1 + \tan^2\theta = \sec^2\theta$ or $\tan^2\theta = \sec^2\theta - 1$

If you answered correctly, go to problem 73.
If you answered incorrectly, review "Pythagorean Identities" on page 140.

73. State the difference identity for sine.

Answer: $\sin(\alpha - \beta) = \sin\alpha\cos\beta - \cos\alpha\sin\beta$

If you answered correctly, go to problem 75.
If you answered incorrectly, go to problem 74.

74. State the sum identity for cosine.

Answer: $\cos(\alpha + \beta) = \cos\alpha\cos\beta - \sin\alpha\sin\beta$

If you answered correctly, go to problem 75.
If you answered incorrectly, review "Addition and Subtraction Identities" on page 143.

75. If M is in the second quadrant and $\sin M = \frac{3}{5}$, find $\sin 2M$.

Answer: $\frac{24}{25}$

If you answered correctly, go to problem 77.
If you answered incorrectly, go to problem 76.

76. If V is a second quadrant angle, and $\sin V = \frac{5}{13}$, what is $\cos 2V$?

Answer: $\frac{119}{169}$

If you answered correctly, go to problem 77.
If you answered incorrectly, review "Double Angle Identities" on page 144.

77. If $\sin\frac{1}{2}\phi = \frac{1}{4}$ and ϕ is an acute angle, find $\cos\phi$.

Answer: $\frac{7}{8}$

If you answered correctly, go to problem 79.
If you answered incorrectly, go to problem 78.

78. If $\cos\phi = -\dfrac{3}{8}$ and ϕ is in Quadrant III, find $\cos\dfrac{1}{2}\phi$.

Answer: $-\dfrac{\sqrt{5}}{4}$

If you answered correctly, go to problem 79.
If you answered incorrectly, review "Half Angle Identities" on page 146.

79. If $\tan\alpha = 8$ and $\tan\beta = \dfrac{3}{4}$ what is the tangent of their sum?

Answer: $-\dfrac{7}{4}$

If you answered correctly, go to problem 81.
If you answered incorrectly, go to problem 80.

80. If $\sigma = \arcsin\dfrac{3}{5}$, find $\tan 2\sigma$.

Answer: $\dfrac{24}{7}$ or $3\dfrac{3}{7}$

If you answered correctly, go to problem 81.
If you answered incorrectly, review "Tangent Identities" on page 150.

81. Write $\cos 3w\cos 2w$ as a sum.

Answer: $\dfrac{\cos 5w}{2} + \dfrac{\cos w}{2}$

If you answered correctly, go to problem 83.
If you answered incorrectly, go to problem 82.

82. Express $\sin 5y\cos 3y$ as a sum.

Answer: $\sin 4y + \sin y$

If you answered correctly, go to problem 83.
If you answered incorrectly, review "Product-Sum Identities" on page 152.

83. Write the difference $\cos 75° - \cos 45°$ as a product.

Answer: $-2\sin 60° \sin 15°$

If you answered correctly, go to problem 85.
If you answered incorrectly, go to problem 84.

84. Write the sum $\sin 5p + \sin 9p$ as a product.

Answer: $2\cos 7p \cos - 2p$

If you answered correctly, go to problem 85.
If you answered incorrectly, review "Sum-Product Identities" on page 153.

85. Give two examples of vectors.

Answer: Answers will vary. They must be examples that have magnitude and direction.

If you answered correctly, go to problem 87.
If you answered incorrectly, go to problem 86.

86. Tell whether and why speed is a vector or scalar.

Answer: scalar; It has magnitude but no direction.

If you answered correctly, go to problem 87.
If you answered incorrectly, review "Vectors versus Scalars" on page 167.

87. What is the Tip-Tail Rule?

Answer: The tip of one component vector is placed next to the tail of the other to form a triangle.

If you answered correctly, go to problem 89.
If you answered incorrectly, go to problem 88.

88. What name is given to the side formed by connecting the tail of one component vector to the tip of the second one?

Answer: the resultant

If you answered correctly, go to problem 89.
If you answered incorrectly, review "Vector Addition Triangle/The Tip-Tail Rule" on page 169.

89. What method of resolving vectors is more popular with scientists than the vector triangle?

Answer: the vector parallelogram or parallelogram of forces

If you answered correctly, go to problem 91.
If you answered incorrectly, go to problem 90.

90. An object is being acted upon by two forces, one of which is 15 lbs and the other 20 lbs. They act with an angle between them of 60°. Find the resultant force and angle to the nearest integer and degree.

Answer: 30 lbs at 43°

If you answered correctly, go to problem 91.
If you answered incorrectly, review "Parallelogram of Forces" on page 172.

91. If the endpoints of vector $\overrightarrow{AB}$ are A (−4, −6) and B (5, 8), what are the coordinates of standard vector $\overrightarrow{OP}$ if $\overrightarrow{OP} = \overrightarrow{AB}$?

Answer: (9, 14)

If you answered correctly, go to problem 93.
If you answered incorrectly, go to problem 92.

92. If the endpoints of vector $\overrightarrow{AB}$ are A (3, −5) and B (11, -9), what are the coordinates of standard vector $\overrightarrow{OP}$ if $\overrightarrow{OP} = \overrightarrow{AB}$?

Answer: (8, −4)

If you answered correctly, go to problem 93.
If you answered incorrectly, review "Vectors in the Rectangular Coordinate System" on page 178.

93. The magnitude of the horizontal component of a force is twice the magnitude of the vertical component. Find the angle the resultant force makes with the larger component.

Answer: $\approx 27°$

If you answered correctly, go to problem 95.
If you answered incorrectly, go to problem 94.

94. Find the rectangular components of the vector $16 \angle 30°$ to the nearest integer.

Answer: $x \approx 14$, $y = 8$

If you answered correctly, go to problem 95.
If you answered incorrectly, review "Resolution of Vectors" on page 180.

95. Vector $\mathbf{f} = (5,9)$; vector $\mathbf{g} = (a,b)$, $\mathbf{f} = \mathbf{g}$. Find the value of b.

Answer: 9

If you answered correctly, go to problem 97.
If you answered incorrectly, go to problem 96.

96. Find the magnitude of vector $\mathbf{q} = (-18,-24)$

Answer: 30

If you answered correctly, go to problem 97.
If you answered incorrectly, review "Algebraic Addition of Vectors" on page 183.

97. If $\mathbf{r} = (c,d)$ and $\mathbf{s} = (g,h)$, find the product of 11 and $\mathbf{r} + \mathbf{s}$.

Answer: $(11c, 11d) + (11g, 11h)$

If you answered correctly, go to problem 99.
If you answered incorrectly, go to problem 98.

98. If $\mathbf{e} = (7, -8)$ and $\mathbf{f} = (-7,12)$, find the product of 6 and $\mathbf{e} - \mathbf{f}$.

Answer: $(42, -84) - (-42, 72)$

If you answered correctly, go to problem 99.
If you answered incorrectly, review "Scalar Multiplication" on page 185.

99. Multiply $\mathbf{v}$ $(-4, 6)$ by $\mathbf{u}$ $(-8, 5)$.

Answer: 62

If you answered correctly, go to problem 101.
If you answered incorrectly, go to problem 100.

100. If $\mathbf{m} = (7, -8)$ and $\mathbf{n} = (-6, 9)$, find the dot product of $\mathbf{m}$ and $\mathbf{n}$.

Answer: -114

If you answered correctly, go to problem 101.
If you answered incorrectly, review "Dot Products" on page 185.

101. What are the polar coordinates of a point with the rectangular coordinates (10, −24)?

Answer: (26, 292.6°)

If you answered correctly, go to problem 103.
If you answered incorrectly, go to problem 102.

102. What are the polar coordinates of a point with the rectangular coordinates (−16, −12)?

Answer: (20, 216.9°)

If you answered correctly, go to problem 103.
If you answered incorrectly, review "Polar Coordinates" on page 203.

103. Convert $P(8, 60°)$ to rectangular coordinates.

Answer: $P(4, 7)$

If you answered correctly, go to problem 105.
If you answered incorrectly, go to problem 104.

104. Convert $P(10, 40°)$ to rectangular coordinates.

Answer: $P(7.7, 6.4)$

If you answered correctly, go to problem 105.
If you answered incorrectly, review "Converting from Polar to Rectangular Coordinates" on page 206.

105. In which quadrant would you plot $-5 + 6i$?

Answer: II

If you answered correctly, go to problem 107.
If you answered incorrectly, go to problem 106.

106. In which quadrant would you plot $-4 + -3i$?

Answer: III

If you answered correctly, go to problem 107.
If you answered incorrectly, review "Plotting Complex Numbers on Rectangular Axes" on page 217.

107. Express the complex number $4 + 4i$ as expressions in the form cisθ.

Answer: $4\sqrt{2}$ (cis45°)

If you answered correctly, go to problem 109.
If you answered incorrectly, go to problem 108.

108. Express the complex number $4 + 4i$ as expressions in the form $r(\cos\theta + i\sin\theta)$.

Answer: $4\sqrt{2}$ (cos45° + $i$sin 45°)

If you answered correctly, go to problem 109.
If you answered incorrectly, review "Plotting Complex Numbers on the Polar Axis" on page 219.

109. Find the difference of $6 + 9i$ and its conjugate.

Answer: $18i$

If you answered correctly, go to problem 110.
If you answered incorrectly, go to problem 111.

110. Find the sum of $12 + 7i$ and its conjugate.

Answer: 24

If you answered correctly, go to problem 112.
If you answered incorrectly, review "Conjugates of Complex Numbers" on page 223.

111. If $q = 8(\cos40° + i\sin40°)$, and $r = 9(\cos60° + i\sin60°)$, then find qr.

Answer: $qr = 72(\cos100° + i\sin100°)$

If you answered correctly, go to problem 113.
If you answered incorrectly, go to problem 112.

112. If $v = 30(\cos80° + i\sin80°)$, $w = 6(\cos120° + i\sin120°)$. Find $v \div w$.

Answer: $5(\cos320° + i\sin320°)$

If you answered correctly, go to problem 113.
If you answered incorrectly, review "Multiplying and Dividing Complex Numbers" on page 224.

113. Square $2 + 2i\sqrt{3}$. Express the answer in polar coordinates.

Answer: $(16, 120°)$

If you answered correctly, go to problem 115.
If you answered incorrectly, go to problem 114.

114. Use De Moivre's Theorem to solve the following: $\left[\sqrt{2}(\cos40° + i\sin40°)\right]^5$. Express the result in polar notation.

Answer: $\left(4\sqrt{2}, 200°\right)$

If you answered correctly, go to problem 115.
If you answered incorrectly, review "Finding Powers of Complex Numbers" on page 225.

115. Find the exact value of $\sin\left(-\frac{1}{2}\right)$

Answer: $30°$ and $\frac{-11}{6}$ and $330°$ or $\frac{11\pi}{6}$

If you answered correctly, go to problem 117.
If you answered incorrectly, go to problem 116.

116. Find the exact value of $\text{Cos}\left(\dfrac{\sqrt{2}}{2}\right)$.

Answer: $45°$ or $\dfrac{\pi}{4}$

If you answered correctly, go to problem 117.
If you answered incorrectly, review "Inverse Sine and Cosine" on page 244.

117. Find the exact value of $\text{Tan}^{-1}\left(-\dfrac{1}{\sqrt{3}}\right)$.

Answer: $330°$ or $\dfrac{11\pi}{6}$

If you answered correctly, go to problem 119.
If you answered incorrectly, go to problem 118.

118. Find the exact value of $\text{Tan}^{-1}(\sqrt{3})$.

Answer: $60°$ or $\dfrac{\pi}{3}$ and $240°$ or $\dfrac{4\pi}{3}$

If you answered correctly, go to problem 119.
If you answered incorrectly, review "Inverse Tangent" on page 249.

119. Without using a calculator or table of trigonometric functions, determine the precise value of $\cos(\tan^{-1} -8)$.

Answer: $\dfrac{\pm\sqrt{65}}{65}$

If you answered correctly, go to problem 121.
If you answered incorrectly, go to problem 120.

120. Without using a calculator or table of trigonometric functions, determine the precise value of $\sin[\sec^{-1}(6)]$.

Answer: $\dfrac{\pm\sqrt{35}}{6}$

If you answered correctly, go to problem 121.
If you answered incorrectly, review "Inverses of Reciprocal Functions" on page 253.

121. Solve for x's primary values: $\sin^2 x \sec x + 2\sin^2 x - \sec x - 2 = 0$

Answer: $90°, 120°, 240°, 270°$

If you answered correctly, go to problem 123.
If you answered incorrectly, go to problem 122.

122. Solve for all x's values: $\cos x - \sqrt{2}\,\sin x = 1$.

Answer: $0° + (360°n), 109.5° + (360°n), 250.5 + (360°n)$

If you answered correctly, go to problem 123.
If you answered incorrectly, review "Trigonometric Equations" on page 257.

123. Find the length of the arc intercepted by a central angle of 150° in a circle of radius 8 cm to the nearest cm. Use 3.14 for the value of π.

Answer: 21 cm

If you answered correctly, go to problem 125.
If you answered incorrectly, go to problem 124.

124. If a point revolves around a circle of radius 16 at a constant rate of 6 revolutions every 2 minutes, what is its angular velocity?

Answer: 18.84 radians per minute

If you answered correctly, go to problem 125.
If you answered incorrectly, review "Uniform Circular Motion" on page 261.

125. The displacement of a spring is given by the equation $d = \cos\alpha t$ where A, the initial displacement is 10 inches, $t = 2\pi$, and $\alpha = 8$.

Answer: 10 inches

If you answered correctly, go to problem 126.
If you answered incorrectly, go to problem 126.

126. The displacement of the end of a pendulum is given by the equation $d = k\sin 2\pi t$. Find the displacement to the nearest centimeter if $t = 5$ seconds and $k = -9$.

Answer: 7.8 cm

If you answered correctly, you are finished. Congratulations!
If you answered incorrectly, review "Simple Harmonic Motion" on page 263.

Appendix A
Summary of Formulas

Basic Trigonometric Functions

$$\sin \alpha = \frac{\text{opposite}}{\text{hypotenuse}}$$

$$\cos \alpha = \frac{\text{adjacent}}{\text{hypotenuse}}$$

$$\tan \alpha = \frac{\text{opposite}}{\text{adjacent}}$$

$$\csc \alpha = \frac{\text{hypotenuse}}{\text{opposite}}$$

$$\sec \alpha = \frac{\text{hypotenuse}}{\text{adjacent}}$$

$$\cot \alpha = \frac{\text{adjacent}}{\text{opposite}}$$

Reciprocal Identities

$$\frac{1}{\sin \theta} = \csc \theta$$

$$\frac{1}{\cos \theta} = \sec \theta$$

$$\frac{1}{\tan \theta} = \cot \theta$$

$$\frac{1}{\cot \theta} = \tan \theta$$

$$\frac{1}{\sec \theta} = \cos \theta$$

$$\frac{1}{\csc \theta} = \sin \theta$$

Ratio Identities

$$\tan\theta = \frac{\sin\theta}{\cos\theta}$$

$$\cot\theta = \frac{\cos\theta}{\sin\theta}$$

Trigonometric Cofunctions

$$\sin A = \frac{a}{c} = \cos B \qquad\qquad \sin B = \frac{b}{c} = \cos A$$

$$\sec A = \frac{c}{b} = \csc B \qquad\qquad \sec B = \frac{c}{a} = \csc A$$

$$\tan A = \frac{a}{b} = \cot B \qquad\qquad \tan B = \frac{b}{a} = \cot A$$

$$\sin\theta = \cos(90° - \theta) \qquad \cos\theta = \sin(90° - \theta)$$

$$\sec\theta = \csc(90° - \theta) \qquad \csc\theta = \sec(90° - \theta)$$

$$\tan\theta = \cot(90° - \theta) \qquad \cot\theta = \tan(90° - \theta)$$

Identities for Negatives

$$\sin(-\theta) = -\sin\theta$$

$$\cos(-\theta) = -\cos\theta$$

$$\tan(-\theta) = -\tan\theta$$

$$\sec(-\theta) = -\sec\theta$$

$$\csc(-\theta) = -\csc\theta$$

$$\cot(-\theta) = -\cot\theta$$

Pythagorean Identities

$$\sin^2\theta + \cos^2\theta = 1$$

$$\sin^2\theta = 1 - \cos^2\theta$$

$$\cos^2\theta = 1 - \sin^2\theta$$

$$\csc^2\theta - \cot^2\theta = 1$$

$$1 + \cot^2\theta = \csc^2\theta$$

$$\cot^2\theta = \csc^2\theta - 1$$

$$\sec^2\theta - \tan^2\theta = 1$$

$$1 + \tan^2\theta = \sec^2\theta$$

$$\tan^2\theta = \sec^2\theta - 1$$

Opposite Angle Identities

$\sin(-\theta) = -\sin\theta$

$\cos(-\theta) = -\cos\theta$

$\tan(-\theta) = -\tan\theta$

Double Angle Identities

$\sin2\theta = 2\sin\theta\cos\theta$

$\cos2\theta = \cos^2\theta - \sin^2\theta$

$\cos2\theta = 2\cos^2\theta - 1$

$\cos2\theta = 1 - 2\sin^2\theta$

$\tan2\theta = \dfrac{2\tan\theta}{1 - \tan^2\theta}$

Half Angle Identities

$\sin\dfrac{\theta}{2} = \pm\sqrt{\dfrac{1 - \cos\theta}{2}}$

$\cos\dfrac{\theta}{2} = \pm\sqrt{\dfrac{1 + \cos\theta}{2}}$

$\tan\dfrac{\theta}{2} = \dfrac{1 - \cos\theta}{\sin\theta}$

$\tan\dfrac{\theta}{2} = \dfrac{\sin\theta}{1 + \cos\theta}$

$\tan\dfrac{\theta}{2} = \pm\sqrt{\dfrac{1 - \cos\theta}{1 + \cos\theta}}$

Sum and Difference Identities

$\sin(\alpha + \beta) = \sin\alpha\cos\beta + \cos\alpha\sin\beta$

$\sin(\alpha - \beta) = \sin\alpha\cos\beta - \cos\alpha\sin\beta$

$\cos(\alpha + \beta) = \cos\alpha\cos\beta - \sin\alpha\sin\beta$

$\cos(\alpha - \beta) = \cos\alpha\cos\beta + \sin\alpha\sin\beta$

$\tan(\alpha + \beta) = \dfrac{\tan\alpha + \tan\beta}{1 - \tan\alpha\tan\beta}$

$\tan(\alpha - \beta) = \dfrac{\tan\alpha - \tan\beta}{1 + \tan\alpha\tan\beta}$

Product-Sum Identities

$$\sin\alpha\cos\beta = \tfrac{1}{2}\left[\sin(\alpha+\beta)+\sin(\alpha-\beta)\right]$$

$$\cos\alpha\sin\beta = \tfrac{1}{2}\left[\sin(\alpha+\beta)-\sin(\alpha-\beta)\right]$$

$$\sin\alpha\sin\beta = \tfrac{1}{2}\left[\cos(\alpha-\beta)-\cos(\alpha+\beta)\right]$$

$$\cos\alpha\cos\beta = \tfrac{1}{2}\left[\cos(\alpha+\beta)+\cos(\alpha-\beta)\right]$$

Sum-Product Identities

$$\sin\alpha + \sin\beta = 2\sin\frac{\alpha+\beta}{2}\cos\frac{\alpha-\beta}{2}$$

$$\sin\alpha - \sin\beta = 2\cos\frac{\alpha+\beta}{2}\cos\frac{\alpha-\beta}{2}$$

$$\cos\alpha + \cos\beta = 2\cos\frac{\alpha+\beta}{2}\cos\frac{\alpha-\beta}{2}$$

$$\cos\alpha - \cos\beta = -2\sin\frac{\alpha+\beta}{2}\sin\frac{\alpha-\beta}{2}$$

Inverse Identities

$$\sin(\text{Sin}^{-1}x) = x \qquad -1 \le x \le 1$$

$$\text{Sin}^{-1}(\sin x) = x \qquad -\frac{\pi}{2} \le x \le \frac{\pi}{2}$$

$$\cos(\text{Cos}^{-1}x) = x \qquad -1 \le x \le 1$$

$$\text{Cos}^{-1}(\cos x) = x \qquad -\frac{\pi}{2} \le x \le \frac{\pi}{2}$$

$$\tan(\text{Tan}^{-1}x) = x \qquad -\infty \ge x \ge \infty$$

$$\text{Tan}^{-1}(\tan x) = x \qquad -\frac{\pi}{2} \le x \le \frac{\pi}{2}$$

$$\text{Csc}^{-1}x = \text{Sin}^{-1}\frac{1}{x} \text{ when } x \le -1 \text{ or } x \ge 1$$

$$\text{Sec}^{-1}x = \text{Cos}^{-1}\frac{1}{x} \text{ when } x \le -1 \text{ or } x \ge 1$$

$$\text{Cot}^{-1}x = \text{Tan}^{-1}\frac{1}{x} \text{ when } x > 0$$

$$\text{Cot}^{-1}x = \pi + \text{Tan}^{-1}\frac{1}{x} \text{ when } x < 0$$

Appendix B
Trigonometric Functions Table

In order to read the value of a trigonometric ratio for a certain angle, you move one finger down the column that reflects the name of the ratio you're looking for at its top, while moving your eyes down the angle values on the left until you find the one for which you're looking. Then move them across to the proper column. Where your eyes and finger meet you'll find the value you want. The tables show values for all six trigonometric functions in increments of 1°.

θ	sin θ	cos θ	tan θ	cot θ	sec θ	csc θ
0°	.000	1.000	.000	Undefined	1.000	Undefined
1°	.017	1.000	.017	57.290	1.000	57.299
2°	.035	.999	.035	28.636	1.001	28.654
3°	.052	.999	.052	19.081	1.001	19.107
4°	.070	.998	.070	14.301	1.002	14.336
5°	.087	.996	.087	11.430	1.004	11.474
6°	.105	.995	.105	9.514	1.006	9.567
7°	.122	.993	.123	8.144	1.008	8.206
8°	.139	.990	.141	7.115	1.010	7.185
9°	.156	.988	.158	6.314	1.012	6.392
10°	.174	.985	.176	5.671	1.015	5.759
11°	.191	.982	.194	5.145	1.019	5.241
12°	.208	.978	.213	4.705	1.022	4.810
13°	.225	.974	.231	4.331	1.026	4.445
14°	.242	.970	.249	4.011	1.031	4.134
15°	.259	.966	.268	3.732	1.035	3.864
16°	.276	.961	.287	3.487	1.040	3.628
17°	.292	.956	.306	3.271	1.046	3.420

(continued)

θ	sin θ	cos θ	tan θ	cot θ	sec θ	csc θ
18°	.309	.951	.325	3.078	1.051	3.236
19°	.326	.946	.344	2.904	1.058	3.072
20°	.342	.940	.364	2.747	1.064	2.924
21°	.358	.934	.384	2.605	1.071	2.790
22°	.375	.927	.404	2.475	1.079	2.669
23°	.391	.921	.424	2.356	1.086	2.559
24°	.407	.914	.445	2.246	1.095	2.459
25°	.423	.906	.466	2.145	1.103	2.366
26°	.438	.899	.488	2.050	1.113	2.281
27°	.454	.891	.510	1.963	1.122	2.203
28°	.469	.883	.532	1.881	1.133	2.130
29°	.485	.875	.554	1.804	1.143	2.063
30°	.500	.866	.577	1.732	1.155	2.000
31°	.515	.857	.601	1.664	1.167	1.972
32°	.530	.848	.625	1.600	1.179	1.887
33°	.545	.839	.649	1.540	1.192	1.836
34°	.559	.829	.675	1.483	1.206	1.788
35°	.574	.819	.700	1.428	1.221	1.743
36°	.588	.809	.727	1.376	1.236	1.701
37°	.602	.799	.754	1.327	1.252	1.662
38°	.616	.788	.781	1.280	1.269	1.624
39°	.629	.777	.810	1.235	1.287	1.589
40°	.643	.766	.839	1.192	1.305	1.556
41°	.656	.755	.869	1.150	1.325	1.524
42°	.669	.743	.900	1.111	1.346	1.494
43°	.682	.731	.933	1.072	1.367	1.466
44°	.695	.719	.966	1.036	1.390	1.440
45°	.707	.707	1.000	1.000	1.414	1.414
46°	.719	.695	1.036	.966	1.440	1.390
47°	.731	.682	1.072	.933	1.466	1.367
48°	.743	.669	1.111	.900	1.494	1.346

θ	$sin\ \theta$	$cos\ \theta$	$tan\ \theta$	$cot\ \theta$	$sec\ \theta$	$csc\ \theta$
49°	.755	.656	1.150	.869	1.524	1.325
50°	.766	.643	1.192	.839	1.556	1.305
51°	.777	.629	1.235	.810	1.589	1.287
52°	.788	.616	1.280	.781	1.624	1.269
53°	.799	.602	1.327	.754	1.662	1.252
54°	.809	.588	1.376	.727	1.701	1.236
55°	.819	.574	1.428	.700	1.743	1.221
56°	.829	.559	1.483	.675	1.788	1.206
57°	.839	.545	1.540	.649	1.836	1.192
58°	.848	.530	1.600	.625	1.887	1.179
59°	.857	.515	1.664	.601	1.972	1.167
60°	.866	.500	1.732	.577	2.000	1.155
61°	.875	.485	1.804	.554	2.063	1.143
62°	.883	.469	1.881	.532	2.130	1.133
63°	.891	.454	1.963	.510	2.203	1.122
64°	.899	.438	2.050	.488	2.281	1.113
65°	.906	.423	2.145	.466	2.366	1.103
66°	.914	.407	2.246	.445	2.459	1.095
67°	.921	.391	2.356	.424	2.559	1.086
68°	.927	.375	2.475	.404	2.669	1.079
69°	.934	.358	2.605	.384	2.790	1.071
70°	.940	.342	2.747	.364	2.924	1.064
71°	.946	.326	2.904	.344	3.072	1.058
72°	.951	.309	3.078	.325	3.236	1.051
73°	.956	.292	3.271	.306	3.420	1.046
74°	.961	.276	3.487	.287	3.628	1.040
75°	.966	.259	3.732	.268	3.864	1.035
76°	.970	.242	4.011	.249	4.134	1.031
77°	.974	.225	4.331	.231	4.445	1.026
78°	.978	.208	4.705	.213	4.810	1.022

(continued)

θ	$sin\ \theta$	$cos\ \theta$	$tan\ \theta$	$cot\ \theta$	$sec\ \theta$	$csc\ \theta$
79°	.982	.191	5.145	.194	5.241	1.019
80°	.985	.174	5.671	.176	5.759	1.015
81°	.988	.156	6.314	.158	6.392	1.012
82°	.990	.139	7.115	.141	7.185	1.010
83°	.993	.122	8.144	.123	8.206	1.008
84°	.995	.105	9.514	.105	9.567	1.006
85°	.996	.087	11.430	.087	11.474	1.004
86°	.998	.070	14.301	.070	14.336	1.002
87°	.999	.052	19.081	.052	19.107	1.001
88°	.999	.035	28.636	.035	28.654	1.001
89°	1.000	.017	57.290	.017	57.299	1.000
90°	1.000	.000	Undefined	.000	Undefined	1.000

Glossary

The number(s) in parentheses indicate the chapter(s) where the term is important.

AAS (3)   A way of referring to a triangle about which you know two angles and a side not included between those angles. Also referred to as SAA.

abscissa (2)   The x-coordinate.

absolute value of a complex number (6)   The square root of the sum of the squares of its real and imaginary parts.

algebraic vector (5)   An ordered pair naming the terminal point of a vector.

amplitude (2)   Vertical displacement of a wave or curve above the x-axis.

angle of depression (3)   The downward angle formed below a horizontal line of sight with that line of sight.

angle of elevation (3)   The upward angle formed above a horizontal line of sight with that line of sight.

angular velocity (7)   Velocity defined in terms of central angle of rotation and time.

area (3)   The region inside a plane figure.

ASA (3)   A way of referring to a triangle about which you know angles and a side included between those angles.

asymptote (2)   A value that a function may approach but never reach; a value that is undefined for a function.

cardoid (6)   A heart-shaped graph created by using polar coordinates.

Cartesian coordinates (5)   Rectilinear coordinate system, invented by French philosopher/mathematician Rene Descartes, which permits location of all points in a plane by assigning them x and y coordinates.

circular functions (2)   Alternative definition of trigonometric functions based upon a unit circle.

complex plane (6)   A plane for graphing complex numbers with the horizontal coordinates being the real part and the vertical being the imaginary.

component vector (5)   One of two or more vectors acting on a single object, point, and so on.

conditional equation (4)   An equation that is true for only certain replacement values.

conjugate of a complex number (6)   The same as the original complex number with a different sign.

coterminal angles (1)   All angles 360° apart that share a common terminal side.

DeMoivre's theorem (6)   A theorem for computing the values of complex numbers raised to any specific exponent.

domain (2)   The horizontal coordinates or inputs.

dot product (5)   The process of combining two vectors, which yields a single number.

degree (2)   One of the 360 equal divisions of the central angle of a circle; each degree may be subdivided into 60 minutes; and each minute may be further subdivided into 60 seconds.

equivalent vectors (5)   Two or more vectors having the same magnitude and direction.

frequency (2)   The occurrence of a complete period per unit of time.

general triangle (3)   Any triangle that is not a right triangle; also oblique triangle.

Heron's formula (3)   A formula for determining the area of any triangle, knowing only its perimeter or the length of its sides.

identity (4, 6)   See *trigonometric identity*.

imaginary axis (6)   The vertical axis in the complex plane.

interpolate (1)   To work out intermediate values for a trigonometric function from known values on either side of an angle in question.

Law of Cosines (3)   A relationship between all three sides of a triangle and the cosine of one.

Law of Sines (3)   A relationship between any two sides of a triangle with the sines of the angle opposite them.

linear velocity (7)   Velocity defined in terms of arc length and time.

negative angle (1, 6)   Any angle resulting from a clockwise rotation of the terminal side.

ordinate (2)   The *y*-coordinate.

parallelogram of vectors (also parallelogram of forces) (5)   A figure made of vectors and used to resolve the action of two vectors on a single object.

period (2)   The smallest occurrence of a complete function, graph, or wave.

periodic functions (2, 6)   Trigonometric functions whose values repeat over regular intervals.

phase shift (2)   The horizontal displacement of a function to the left or right; the cosine graph is 90° out of phase with the sine function.

polar axis (6)   A horizontal ray extending from the pole to the right, which marks the reference ray from which angle measures begin in the polar coordinate system.

polar coordinates (6)   An ordered pair consisting of a distance from the pole and an angle.

polar coordinate system (6)   A coordinate system that relies upon distance and angle from the pole to determine position.

pole (6)   The origin in polar coordinates.

Pythagorean identities (4)   Fundamental identities relating to sine and cosine functions and the Pythagorean theorem.

Pythagorean theorem (4)   The relationship that the square on the hypotenuse of a right triangle is equal to the sum of the squares on the legs.

quadrant (1, 2, 3, 4, 5, 6, 7)   One of the four partitions formed by the crossing of the *x* and *y* axes and designated by Roman numerals I–IV.

quadrantal angle (1)   An angle that in standard position has its terminal side fall on one of the axes.

radian (2)   A measure of angle size relating a central angle of a circle to an arc equal in length to a radius of the circle; 2π radians are equivalent to $360°$.

range (2)   The vertical coordinates or outputs.

ratio identity (4)   The identities relating tangent and cotangent to the sine and cosine functions.

real axis (6)   The horizontal axis in the complex plane.

reciprocal identities (4)   Identities formed by placing a one over each of the basic trigonometric functions.

reference angle (1)   The acute angle in nonstandard position that is used to find the trigonometric function for angles greater than $90°$.

resultant (5)   The product of a vector multiplication.

SAA (3)   A way of referring to a triangle about which you know two angles and a side not included between those angles. Also referred to as AAS.

SSA (3)   A way of referring to a triangle about which you know two sides and an angle not included between those sides. This is also known as the ambiguous case.

SAS (3)   A way of referring to a triangle about which you know two sides and an angle included between those sides.

scalar (5)   A quantity with magnitude only (as distinguished from vector).

simple harmonic motion (7)   A component of uniform circular motion and characteristic of springs, pendulums, and pistons.

SOHCAHTOA (1)   A mnemonic device for remembering the three basic trigonometric functions.

SSS (3)   A way of referring to a triangle about which you know all three sides, but no angle.

standard position angle (1)   An angle with its initial side on the horizontal axis right of the origin.

tip-tail rule (5)   A method for performing vector addition.

trigonometric function (1, 7)   Any of the six relationships possible between two sides of a right triangle.

trigonometric identity (1, 4, 7)   An equation relating trigonometric functions that is true for all input values.

uniform circular motion (7)   Motion at a fixed distance about a single point at a uniform linear and angular velocity.

unit circle (2)   A circle with radius 1.

vector (5)   A quantity with both magnitude and direction.

vector resolution (5)   Resolving an oblique vector into its horizontal and vertical components.

wavelength (2)   The distance between crests on a periodic graph or wave.

zero vector (5)   A vector with zero magnitude that points in any direction.

Index

Symbols and Numerics

α (alpha), angle measure designated by, 28

$\approx$ (approximately equals sign), 43

β (beta), angle measure designated by, 28

first quadrant angle, 21

four-leaved rose polar graphs, 216–217

fourth quadrant angle, 23

one

    reciprocals and function value of, 81

    square root of negative (i), 217

φ (phi), angle measure designated by, 28

π (pi)

    periodic functions and, 66–67

    radian measurement and, 60

second quadrant angle, 22

θ (theta), angle measure designated by, 28

third quadrant angle, 24

30-60-90 right triangle ratios (table), 36

three-leaved rose polar graphs, 215–216

zero

    reciprocals and function value

      approaching, 81

    zero vector, 183, 303

A

AAS (Angle-Angle-Side) triangles

    defined, 301, 303

    finding the area, 117

    solving the triangle, 111

abscissa (x-value)

    defined, 62, 301

    domain and, 63

    function definition and, 62, 241

absolute value

    of angles, 22

    of complex numbers, 219, 301

ACTS mnemonic for positive ratios, 40

acute reference angles, 41–42, 303

addition identities

    for cosine, 143, 295

    for sine, 143, 295

    for tangent, 150, 295

addition of vectors

    algebraic, 183

    components, 169, 183

    parallelogram of forces for, 172–173

    resultant, 169, 173

    tip-tail rule, 169, 303

    triangle representing, 169

adjacent side

    defined, 28

    in reciprocal functions, 30

    in trigonometric function definitions, 28, 293

algebraic addition of vectors, 183

algebraic vectors, defined, 183, 301

alpha (α), angle measure designated by, 28

ambiguous case, 111–112

amplitude

    of complex numbers, 219

    cosine function amplitude shifts, 71

    cotangent function and, 79

    defined, 71, 301

    sine function amplitude shifts, 71

    tangent function and, 79

angle measure, 21

angle of depression, 96, 301

angle of elevation, 95, 301

Angle-Angle-Side (AAS) triangles

    defined, 301, 303

    finding the area, 117

    solving the triangle, 111

angles

    absolute value, 22

    coterminal, 25–26, 301

    defined, 21

    of depression, 96, 301

    double angle identities, 144, 150, 295

    of elevation, 95, 301

    half angle identities, 146, 150, 295

    initial side, 21

    negative, 302

    quadrantal, 23, 40, 303

    reference, 41–42, 303

    standard position, 21, 303

    terminal side, 21

Angle-Side-Angle (ASA) triangles
  defined, 301
  finding the area, 117
  solving the triangle, 111
angular velocity, 262, 301
approximately equals sign ($\approx$), 43
arccos function on scientific calculators, 45
Archimedes Spiral, 211
arcsin function on scientific calculators, 45
arctan function on scientific calculators, 45
area, defined, 301
area of triangles
  common formula for, 116
  formulas for ASA or SAA triangles, 117
  formulas for SAS triangles, 116
  Heron's formula for SSS triangles, 119,
    302
argument of complex number, 219
arrows, vectors represented by, 167
ASA (Angle-Side-Angle) triangles
  defined, 301
  finding the area, 117
  solving the triangle, 111
asymptotes
  cosecant function, 81
  cotangent function, 79
  defined, 77, 301
  secant function, 81
  solving for, 79
  tangent function, 77

B

beta (β), angle measure designated by, 28
boldface for vectors, 167

C

calculators, scientific, 45
cardoid
  defined, 301
  polar graphs, 213–214
Cartesian coordinate plane. *See also*
 quadrants
  complex plane versus, 217–218
  converting from polar to rectangular
    coordinates, 206

  converting from rectangular to polar
    coordinates, 206
  defined, 301
  resolution of forces, 180
  vectors in, 178–179
CAST mnemonic for positive ratios, 40
centered vector, 179
circle
  arc subtended by 1 degree, 59
  arc subtended by 1 radian, 59–60
  polar graphs of, 211–212
  radian measurement and, 60
  unit circle and circular functions, 62–63
  unit circle and periodic functions, 66–67
  unit circle, defined, 303
circular functions
  defined, 62, 301
  domain versus range of, 63
  periodic properties of, 66–67
circular motion, uniform
  angular velocity, 262
  central angle, 261
  defined, 303
  linear velocity, 262
  negative velocities, 262
  simple harmonic motion and, 261, 263
cofunction identities, 139, 294
cofunctions, defined, 34, 294. *See also*
 specific cofunctions
collecting terms, 258
commutative property of multiplication, 185
complex numbers
  absolute value, 219, 301
  amplitude, 219
  argument, 219
  conjugates of, 223, 301
  DeMoivre's theorem, 225, 301
  difference of two conjugates, 223
  dividing, 224
  equal, defined, 223
  finding powers of, 225
  modulus, 219
  multiplying, 224
  plotting on polar axis, 219–220
  plotting on rectangular axes, 217–218
  polar form, 219
  sum of two conjugates, 223
  trigonometric form, 219

complex plane, 217–218, 301
component vector, 301
components of vector addition, 169, 183
conditional equations
    collecting terms, 258
    defined, 137, 257, 301
    general solution, 257
    primary solutions, 257
    solving, 258
conjugates of complex numbers, 223, 301
converting. *See also* phase shift
    from polar to rectangular coordinates, 206
    from rectangular to polar coordinates, 206
    sine curve to and from cosine curve, 76
coordinate systems. *See* polar coordinate system; rectangular coordinate system
cosecant (csc)
    asymptotes, 81
    circular function, 62
    cofunction, 34, 294
    cofunction identities, 139
    defined, 29–30, 293
    finding with scientific calculators, 45
    graphing sine on same axes, 80
    identity for inverse cosecant, 253, 296
    identity for negatives, 139, 294
    inverse function, 253
    period, 81
    phase shift between secant and, 81
    Pythagorean identities, 140, 294
    for quadrantal angles (table), 40
    reciprocal identity, 137, 293
    sine as reciprocal function, 29–30
    for special triangles (table), 36
    values in increments of one degree (table), 297–300
cosine (cos)
    as basic circular function, 62
    cofunction, 34, 294
    cofunction identities, 139
    defined, 28, 293
    difference identity, 143, 295
    double angle identities, 144, 295
    finding with scientific calculators, 45
    graphing amplitude shifts, 71

graphing one period, 69
graphing phase shifts, 76
graphing sine on same axes, 69
graphing various frequencies, 75
graphing vertical shifts, 70
half angle identity, 146, 295
identities for inverse cosine, 247, 296
identity for negatives, 139, 294
inverse function, 245–247
Law of Cosines, 105, 302
opposite angle identity, 295
as periodic function, 66–67
phase shift between sine and, 69, 75
product-sum identities, 152, 296
Pythagorean identities, 140, 294
for quadrantal angles (table), 40
ratio identities, 138, 294
reciprocal identity, 137, 293
restricted function, 246–247
restricted inverse function, 246–247
secant as reciprocal function, 29–30
for special triangles (table), 36
sum identity, 143, 295
sum-product identities, 153, 296
trigonometric identities, 32–33
turning into a sine, 76
values for points in one period (table), 68
values in increments of one degree (table), 297–300
cotangent (cot)
    asymptotes, 79
    circular function, 62
    cofunction, 34, 294
    cofunction identities, 139
    defined, 29–30, 293
    graphing, 78–79
    identities for inverse cotangent, 255, 296
    identity for negatives, 139, 294
    inverse function, 255
    period, 79
    phase shift, 79
    Pythagorean identities, 140, 294
    for quadrantal angles (table), 40
    ratio identity, 138, 294
    reciprocal identity, 137, 293
    for special triangles (table), 36
    tangent as reciprocal function, 29–30

cotangent (cot) *(continued)*
  trigonometric identity, 32
  values in increments of one degree (table),
    297–300
coterminal angles
  defined, 25, 301
  equation for, 26
counterexamples, 137
csc (cosecant)
  asymptotes, 81
  circular function, 62
  cofunction, 34, 294
  cofunction identities, 139
  defined, 29–30, 293
  finding with scientific calculators, 45
  graphing sine on same axes, 80
  identity for inverse cosecant, 253, 296
  identity for negatives, 139, 294
  inverse function, 253
  period, 81
  phase shift between secant and, 81
  Pythagorean identities, 140, 294
  for quadrantal angles (table), 40
  reciprocal identity, 137, 293
  sine as reciprocal function, 29–30
  for special triangles (table), 36
  values in increments of one degree (table),
    297–300
customized full-length exam, 275–292

D

degrees. *See also* radians
  arcs subtended by, 59
  common values (table), 60
  decimal parts of, 45
  defined, 59, 302
  traditional subdivisions (minutes and
    seconds), 45
DeMoivre, Abraham (mathematician), 225
DeMoivre's theorem, 225, 301
Descartes, Rene (mathematician), 301
difference identities
  for cosine, 143, 295
  for sine, 143, 295
  for tangent, 150, 295
difference of two complex conjugates, 223

direction
  of measure, negative angles and, 22, 23,
    203
  as property of vectors, 168
  of rotation, negative velocity and, 262
dividing complex numbers, 224
domain
  defined, 301
  of functions, 63
  range and, 63
dot product, 185, 302
double angle identities
  for cosine, 144, 295
  for sine, 144, 295
  for tangent, 150, 295

E

electrical engineering notation for vectors,
  180
equality of complex numbers, 223
equals sign, squiggly ($\approx$), 43
equations, trigonometric. *See also* formulas,
  trigonometric
  collecting terms, 258
  conditional, 137, 257–258, 301
  solving, 258
  trigonometric identities, 32
equivalent vectors, 168, 302
exams
  full-length exam, 275–292
  pretest, 1–20

F

first quadrant angle, 21
formulas, trigonometric. *See also* equations,
  trigonometric
  basic trigonometric functions, 28, 30, 293
  cofunction identities, 139, 294
  cofunctions, 34, 294
  cosecant, 30, 293
  cosine, 28, 293
  cotangent, 30, 293
  difference identities, 143, 150, 295
  double angle identities, 144, 150, 295
  half angle identities, 146, 150, 295

Heron's formula, 119, 302
identities for negatives, 139, 294
inverse cosecant, 253, 296
inverse cosine, 247, 296
inverse cotangent, 255, 296
inverse secant, 254, 296
inverse sine, 245, 296
inverse tangent, 251, 296
opposite angle identities, 295
product-sum identities, 152, 296
Pythagorean identities, 140, 294
ratio identities, 138, 294
reciprocal identities, 137, 293
secant, 30, 293
sine, 28, 293
sum identities, 143, 150, 295
sum-product identities, 153, 296
tangent, 28, 293
tangent identities, 150
four-leaved rose polar graphs, 216–217
fourth quadrant angle, 23
free vector, 179
frequency
cosine function with various frequencies, 75
defined, 74, 302
sine function with various frequencies, 74
full-length exam, 275–292
functions. *See also* inverse functions; *specific
functions*
basic functions, 293
circular functions, 62–63, 66–67, 301
cofunction identities, 139, 294
cofunctions, 34, 294
defined, 62, 241, 303
for isosceles right triangle, 35, 36
overview, 28–29
parabola, 241
periodic functions, 66–67, 70, 302
quadrants and, 38–40
reciprocal functions, 29–30, 34, 81, 137
SOHCAHTOA acronym for, 28, 303
table of values, 297–300
for 30-60-90 right triangle, 36
vertical line test for, 241, 242
fundamental identities. *See also* trigonometric
identities
cofunction identities, 139, 294
for negatives, 139, 294

Pythagorean identities, 140, 294, 302
ratio identities, 138, 294, 303
reciprocal identities, 137, 293, 303

G

general solution of conditional equations,
257
general triangles
areas of, 116, 117, 119
defined, 302
solving ASA triangles, 111
solving SAA or AAS triangles, 111
solving SAS triangles, 109
solving SSA triangles (ambiguous case),
111–112
solving SSS triangles, 109
strategies for solving, 108
glossary, 301–303
graphs. *See also* graphs (cosine function);
graphs (sine function)
asymptotes in, 77
circular functions, 62
cosecant function, 80–81
cotangent function, 78–79
parabola, 241
periodic functions, 66
polar graphs, 211–217
principles of reciprocal functions, 81
secant function, 80–81
tangent function, 77–78
unit circle, 62, 66
wavelength, 303
graphs (cosine function)
amplitude shifts, 71
cosine and inverse cosine function,
247–248
frequencies, 75
one period, 69
phase shifts, 76
with secant on same axes, 80
with sine on same axes, 69
vertical shifts, 70
graphs (sine function)
amplitude shifts, 71
circular motion and sine curve, 261
with cosecant on same axes, 80
with cosine on same axes, 69

graphs (sine function) *(continued)*
 frequencies, 74
 one period, 69
 phase shifts, 75
 sine and inverse sine function, 247–248
 vertical and amplitude shifts, 72
 vertical shifts, 70
Greek letters, angle measure designated
 by, 28

H

half angle identities
 for cosine, 146, 295
 for sine, 146, 295
 for tangent, 150, 295
harmonic motion, simple
 defined, 303
 overview, 263
 uniform circular motion and, 261, 263
Heron's formula, 119, 302
horizontal line, polar graph of, 212
hypotenuse
 defined, 28
 of isosceles right triangle, 35
 in reciprocal functions, 30
 in trigonometric function definitions,
  28, 293

I

i (square root of negative one), 217. *See also*
 complex numbers
identities
 cofunction identities, 139
 counterexamples, 137
 defined, 32, 137, 303
 difference identities, 143, 150, 295
 double angle identities, 144, 150, 295
 fundamental identities, 137–140
 half angle identities, 146, 150, 295
 for inverse cosecant, 253, 296
 for inverse cosine, 247, 296
 for inverse cotangent, 255, 296
 for inverse secant, 254, 296
 for inverse sine, 245, 296
 for inverse tangent, 251, 296

for negatives, 139, 294
 opposite angle identities, 295
 product-sum identities, 152, 296
 Pythagorean identities, 140, 294, 302
 ratio identities, 138, 294, 303
 reciprocal identities, 137, 293, 303
 sum identities, 143, 150, 295
 sum-product identities, 153, 296
 tangent identities, 150
imaginary axis, 217, 218, 302
infinity. *See also* asymptotes
 approached by reciprocals as function value
  approaches zero, 81
 approached in tangent graph, 77
initial side of an angle, 21
interpolation, 45–46, 302
inverse functions
 circular functions, 62
 of cosecant function, 253
 of cosine function, 245–247
 of cotangent function, 255
 defined, 242
 of reciprocal functions, 253–255
 restricting, 242–244
 of secant function, 254
 of sine function, 244–245
 symmetry of, 247–248
 of tangent function, 249–251
 vertical line test and, 242
inverse identities
 for cosecant, 253, 296
 for cosine, 247, 296
 for cotangent, 255, 296
 for secant, 254, 296
 for sine, 245, 296
 for tangent, 251, 296
isosceles right triangle
 formula for hypotenuse, 35
 trigonometric ratios (table), 36

L

Law of Cosines
 defined, 105, 302
 Law of Sines versus, 105
 overview, 105
 parallelogram of forces and, 173

solving ASA triangles, 111
solving SAS triangles, 109
solving SSS triangles, 109
Law of Sines
  defined, 100, 302
  Law of Cosines versus, 105
  overview, 100
  solving ASA triangles, 111
  solving SAA or AAS triangles, 111
  solving SAS triangles, 109
  solving SSS triangles, 109
lemniscate polar graphs, 214–215
line of sight
  angle of depression and, 96
  angle of elevation and, 95
linear velocity, 262, 302
lines
  polar graphs of, 212–213
  vertical line test for functions, 241, 242

M

magnitude
  of scalars, 167
  of vectors, 167, 180, 183
  of zero vector, 183, 303
minutes, defined, 45
modulus of complex number, 219
motion
  simple harmonic, 261, 263, 303
  uniform circular, 261–262, 263, 303
multiplying
  complex numbers, 224
  dot product, 185, 302
  vectors by scalars, 185

N

negatives
  angles, defined, 302
  angles, direction of measure and, 22, 23,
    203
  cosine value in quadrant II, 39, 105, 109,
    146
  domain values, 63
  function values in certain quadrants, 39
  identities of, 139, 294

reciprocal functions and, 81
reference angles and, 41
square root of negative one (*i*), 217
velocity, direction of rotation and, 262

O

oblique triangles
  areas of, 116, 117, 119
  defined, 302
  solving ASA triangles, 111
  solving SAA or AAS triangles, 111
  solving SAS triangles, 109
  solving SSA triangles (ambiguous case),
    111–112
  solving SSS triangles, 109
  strategies for solving, 108
oblique vector, 180
obtuse triangles, Law of Sines and, 100
one
  reciprocals and function value of, 81
  square root of negative (*i*), 217
opposite angle identities, 295
opposite side
  in reciprocal functions, 30
  in trigonometric function definitions,
    28, 293
opposite vectors, 168
ordinate (y-value)
  defined, 62, 302
  function definition and, 62, 241
  range and, 63
origin in polar coordinate system, 203, 302

P

parabola, 241
parallelogram of forces, 172–173, 302
period
  of cosecant function, 81
  of cotangent function, 79
  defined, 66, 302
  of secant function, 81
  of tangent function, 77
periodic functions
  amplitude, 70
  defined, 66, 302

periodic functions *(continued)*
overview, 66–67
period defined, 66, 302
real-world applications, 67
wavelength, 303
phase shift
converting sine curve to and from cosine
curve, 76
of cotangent function, 79
defined, 302
graphing cosine with two phase shifts, 76
graphing sine with two phase shifts, 75
between secant and cosecant, 81
between sine and cosine, 69, 75
of tangent function, 79
phi (φ), angle measure designated by, 28
pi (π)
periodic functions and, 66–67
radian measurement and, 60
polar axis, 203, 302
polar coordinate system
converting from polar to rectangular
coordinates, 206
converting from rectangular to polar
coordinates, 206
defined, 203, 302
plotting complex numbers on, 219–220
polar axis, 203, 302
polar coordinates, 203, 302
polar graphs, 211–217
pole or origin, 203, 302
positive versus negative angles in,
203–204
polar coordinates, 203, 302
polar form of complex numbers, 219
pole, 203, 302
position vector, 179
powers of complex numbers, finding, 225
pretest, 1–20
primary solutions of conditional equations,
257
product-sum identities, 152, 296
protractor, 59
Pythagorean identities, 140, 294, 302
Pythagorean theorem
defined, 302
Law of Cosines and, 105
solving the triangle using, 93

Q

quadrantal angles
defined, 23, 303
trigonometric ratios (table), 40
quadrants
defined, 21, 302
first quadrant angle, 21
fourth quadrant negative angle, 23
quadrantal angles, 23
reference angles, 41–42
second quadrant angle, 22
second quadrant negative angle, 22
standard position, 21
third quadrant angle, 24
trigonometric functions and, 38–40
quotient (ratio) identities, 138, 294, 303

R

radians. *See also* degrees
circular functions and, 62
circular motion and, 261–262
common values (table), 60
defined, 60, 303
pi and, 60
radius vector, 179
range, defined, 303
range of functions, 63
ratio identities, 138, 294, 303
rays, vectors versus, 167
real axis, 217, 218, 303
reciprocal functions. *See also specific
functions*
circular functions, 34
inverses of, 253–255
overview, 29–30
principles true of, 81
reciprocal identities and, 137
reciprocal identities, 137, 293, 303
rectangular coordinate system. *See also*
quadrants
complex plane versus, 217–218
converting from polar to rectangular
coordinates, 206
converting from rectangular to polar
coordinates, 206

plotting complex numbers on rectangular
axes, 217–218
resolution of forces, 180
vectors in, 178–179
reference angles, 41–42, 303
resolution of vectors, 180, 303
restricting functions
cosine function, 246
inverse cosine function, 246–247
inverse functions, 242–244
inverse sine function, 245
inverse tangent function, 251
sine function, 244
tangent function, 249–250
resultant
for parallelogram of forces, 173
of vector addition, 169
of vector multiplication, 303

S

SAA (Side-Angle-Angle) triangles
defined, 301, 303
finding the area, 117
solving the triangle, 111
SAS (Side-Angle-Side) triangles
defined, 303
finding the area, 116
solving the triangle, 109
scalar multiplication
as commutative, 185
dot product, 185, 302
multiplication by a scalar versus, 185
scalars
defined, 167, 303
multiplying vectors by, 185
speed as, 167
scientific calculators, 45
secant (sec)
asymptotes, 81
circular function, 62
cofunction, 34, 294
cofunction identities, 139
cosine as reciprocal function, 29–30
defined, 29–30, 293
graphing cosine on same axes, 80
identity for inverse secant, 254, 296

identity for negatives, 139, 294
inverse function, 254
period, 81
phase shift between cosecant and, 81
Pythagorean identities, 140, 294
for quadrantal angles (table), 40
reciprocal identity, 137, 293
for special triangles (table), 36
values in increments of one degree (table),
297–300
second quadrant angle, 22
seconds, defined, 45
Side-Angle-Angle (SAA) triangles
defined, 301, 303
finding the area, 117
solving the triangle, 111
Side-Angle-Side (SAS) triangles
defined, 303
finding the area, 116
solving the triangle, 109
Side-Side-Angle (SSA) triangles
defined, 303
solving the triangle, 111–112
Side-Side-Side (SSS) triangles
defined, 303
finding the area (Heron's formula), 119,
302
solving the triangle, 109
sign
cosine value negative in quadrant II,
39, 105, 109, 146
of function values in certain quadrants, 39
identities of negatives, 139, 294
negative angles and direction of measure,
22, 23, 203
negative angles defined, 302
negative domain values, 63
negative velocity and direction of rotation,
262
reciprocal functions and, 81
reference angles and, 41
square root of negative one (i), 217
similar triangles, 28
simple harmonic motion
defined, 303
overview, 263
uniform circular motion and, 261, 263

sine (sin)
  as basic circular function, 62
  circular motion and sine curve, 261
  cofunction, 34, 294
  cofunction identities, 139
  cosecant reciprocal function, 29–30
  defined, 28, 293
  difference identity, 143, 295
  double angle identity, 144, 295
  finding with scientific calculators, 45
  graphing amplitude shifts, 71
  graphing cosine on same axes, 69
  graphing one period, 69
  graphing phase shifts, 75
  graphing various frequencies, 74
  graphing vertical and amplitude shifts, 72
  graphing vertical shifts, 70
  half angle identity, 146, 295
  identities for inverse sine, 245, 296
  identity for negatives, 139, 294
  inverse function, 244–245
  for isosceles right triangle, 36
  Law of Sines, 100, 302
  opposite angle identity, 295
  as periodic function, 66–67
  phase shift between cosine and, 69, 75
  product-sum identities, 152, 296
  Pythagorean identities, 140, 294
  for quadrantal angles (table), 40
  ratio identities, 138, 294
  reciprocal identity, 137, 293
  restricted function, 244
  restricted inverse function, 245
  for special triangles (table), 36
  sum identity, 143, 295
  sum-product identities, 153, 296
  trigonometric identities, 32–33
  turning into a cosine, 76
  values for points in one period (table), 68
  values in increments of one degree (table),
    297–300
SOHCAHTOA acronym, 28, 303
solving triangles
  ambiguous case, 111–112
  Law of Cosines, 105, 109
  Law of Sines, 100, 109
  oblique triangles, 108–109, 111–112
  right triangles, 93–94

  three sides known (SSS), 109
  two angles and side known (ASA), 111
  two angles and side known (SAA or AAS),
    111
  two sides and angle known (SAS), 109
  two sides and angle known (SSA),
    111–112
square root of negative one (i), 217. *See also*
  complex numbers
squiggly equals sign ($\approx$), 43
SSA (Side-Side-Angle) triangles
  defined, 303
  solving the triangle, 111–112
SSS (Side-Side-Side) triangles
  defined, 303
  finding the area (Heron's formula),
    119, 302
  solving the triangle, 109
standard position of a vector, 178–179
standard position of an angle
  defined, 21, 303
  first quadrant angle in, 21
  second quadrant angle in, 22
standard vector, 179
subtending
  defined, 59
  degree definition and, 59
  radian definition and, 60
subtraction identities
  for cosine, 143, 295
  for sine, 143, 295
  for tangent, 150, 295
sum identities
  for cosine, 143, 295
  for sine, 143, 295
  for tangent, 150, 295
sum of two complex conjugates, 223
sum-product identities, 153, 296
symmetry of inverse functions, 247–248

T

table of degree/radian values, 60
tables of trigonometric ratios
  for 30°, 45°, and 60° angles, 36
  calculators versus, 45
  in increments of one degree, 297–300
  interpolation with, 45–46

for quadrantal angles, 40
sine and cosine for points in one period, 68
using, 44–45
tangent (tan)
asymptotes, 77
circular function, 62
cofunction, 34, 294
cofunction identities, 139
cotangent as reciprocal function, 29–30
defined, 28, 293
difference identity, 150, 295
double angle identity, 150, 295
finding with scientific calculators, 45
graphing, infinity approached in, 77
graphing several cycles, 78
half angle identity, 150, 295
identities for inverse tangent, 251, 296
identity for negatives, 139, 294
inverse function, 249–251
opposite angle identity, 295
period, 77
phase shift, 79
Pythagorean identities, 140, 294
for quadrantal angles (table), 40
ratio identity, 138, 294
reciprocal identity, 137, 293
restricted function, 249–250
restricted inverse function, 251
for special triangles (table), 36
sum identity, 150, 295
trigonometric identity, 32
values in increments of one degree (table),
297–300
terminal side of an angle, 21
terms, collecting, 258
tests
full-length exam, 275–292
pretest, 1–20
theta (θ), angle measure designated by, 28
third quadrant angle, 24
30-60-90 right triangle ratios (table), 36
three-leaved rose polar graphs, 215–216
tip-tail rule, 169, 303
triangle representing vector addition, 169
triangles, finding the area
common formula for, 116
formulas for ASA or SAA triangles, 117

formulas for SAS triangles, 116
Heron's formula for SSS triangles, 119, 302
triangles, solving
ambiguous case, 111–112
Law of Cosines, 105, 109
Law of Sines, 100, 109
oblique triangles, 108–109, 111–112
right triangles, 93–94
three sides known (SSS), 109
two angles and side known (ASA), 111
two angles and side known (SAA or AAS),
111
two sides and angle known (SAS), 109
two sides and angle known (SSA), 111–112
trig tables
for 30°, 45°, and 60° angles, 36
calculators versus, 45
degree/radian values, 60
in increments of one degree, 297–300
interpolation with, 45–46
for quadrantal angles, 40
sine and cosine for points in one period, 68
using, 44–45
trigonometric addition identities
for cosine, 143, 295
for sine, 143, 295
for tangent, 150, 295
trigonometric equations. *See also*
trigonometric formulas
collecting terms, 258
conditional, 137, 257–258, 301
solving, 258
trigonometric identities, 32
trigonometric form of complex numbers, 219
trigonometric formulas. *See also*
trigonometric equations
basic trigonometric functions, 28, 30, 293
cofunction identities, 139, 294
cofunctions, 34, 294
cosecant, 30, 293
cosine, 28, 293
cotangent, 30, 293
difference identities, 143, 150, 295
double angle identities, 144, 150, 295
half angle identities, 146, 150, 295
Heron's formula, 119, 302
identities for negatives, 139, 294

trigonometric formulas *(continued)*
   inverse cosecant, 253, 296
   inverse cosine, 247, 296
   inverse cotangent, 255, 296
   inverse secant, 254, 296
   inverse sine, 245, 296
   inverse tangent, 251, 296
   opposite angle identities, 295
   product-sum identities, 152, 296
   Pythagorean identities, 140, 294
   ratio identities, 138, 294
   reciprocal identities, 137, 293
   secant, 30, 293
   sine, 28, 293
   sum identities, 143, 150, 295
   sum-product identities, 153, 296
   tangent, 28, 293
   tangent identities, 150
trigonometric functions. *See also* inverse
   functions; *specific functions*
   basic functions, 293
   circular functions, 62–63, 301
   cofunction identities, 139
   cofunctions, 34, 294
   defined, 62, 303
   function, defined, 241
   for isosceles right triangle, 35, 36
   overview, 28–29
   periodic functions, 66–67
   quadrants and, 38–40
   reciprocal functions, 29–30, 34, 81, 137
   SOHCAHTOA acronym for, 28, 303
   table of values, 297–300
   for 30-60-90 right triangle, 36
   vertical line test for, 241, 242
trigonometric identities
   cofunction identities, 139
   counterexamples, 137
   defined, 32, 137, 303
   difference identities, 143, 150, 295
   double angle identities, 144, 150, 295
   fundamental identities, 137–140
   half angle identities, 146, 150, 295
   for inverse cosecant, 253, 296
   for inverse cosine, 247, 296
   for inverse cotangent, 255, 296
   for inverse secant, 254, 296
   for inverse sine, 245, 296

   for inverse tangent, 251, 296
   for negatives, 139, 294
   opposite angle identities, 295
   product-sum identities, 152, 296
   Pythagorean identities, 140, 294, 302
   ratio identities, 138, 294, 303
   reciprocal identities, 137, 293, 303
   sum identities, 143, 150, 295
   sum-product identities, 153, 296
   tangent identities, 150
trigonometric tables
   for 30°, 45°, and 60° angles, 36
   calculators versus, 45
   degree/radian values, 60
   in increments of one degree, 297–300
   interpolation with, 45–46
   for quadrantal angles, 40
   sine and cosine for points in one period, 68
   using, 44–45
trigonometry, defined, 21

U

uniform circular motion
   angular velocity, 262
   central angle, 261
   defined, 303
   linear velocity, 262
   negative velocities, 262
   simple harmonic motion and, 261, 263
unit circle
   circular functions and, 62–63
   circumference of, 66
   defined, 62, 303
   periodic functions and, 66–67
   radius of, 62

V

vector resolution, 180, 303
vectors
   addition triangle, 169
   algebraic addition of, 183
   algebraic, defined, 183, 301
   arrows representing, 167
   boldface for, 167
   component vector, defined, 301